T0134762

Studies in Applied Philosophy, Epistemology and Rational Ethics

Volume 50

Studies in Applied Philosophy, Epistemology and Rational Ethics (SAPERE) publishes new developments and advances in all the fields of philosophy, epistemology, and ethics, bringing them together with a cluster of scientific disciplines and technological outcomes: ranging from computer science to life sciences, from economics, law, and education to engineering, logic, and mathematics, from medicine to physics, human sciences, and politics. The series aims at covering all the challenging philosophical and ethical themes of contemporary society, making them appropriately applicable to contemporary theoretical and practical problems, impasses, controversies, and conflicts. Our scientific and technological era has offered "new" topics to all areas of philosophy and ethics – for instance concerning scientific rationality, creativity, human and artificial intelligence, social and folk epistemology, ordinary reasoning, cognitive niches and cultural evolution, ecological crisis, ecologically situated rationality, consciousness, freedom and responsibility, human identity and uniqueness, cooperation, altruism, intersubjectivity and empathy, spirituality, violence. The impact of such topics has been mainly undermined by contemporary cultural settings, whereas they should increase the demand of interdisciplinary applied knowledge and fresh and original understanding. In turn, traditional philosophical and ethical themes have been profoundly affected and transformed as well: they should be further examined as embedded and applied within their scientific and technological environments so to update their received and often old-fashioned disciplinary treatment and appeal. Applying philosophy individuates therefore a new research commitment for the 21st century, focused on the main problems of recent methodological, logical, epistemological, and cognitive aspects of modeling activities employed both in intellectual and scientific discovery, and in technological innovation, including the computational tools intertwined with such practices, to understand them in a wide and integrated perspective.

Studies in Applied Philosophy, Epistemology and Rational Ethics means to demonstrate the contemporary practical relevance of this novel philosophical approach and thus to provide a home for monographs, lecture notes, selected contributions from specialized conferences and workshops as well as selected PhD theses. The series welcomes contributions from philosophers as well as from scientists, engineers, and intellectuals interested in showing how applying philosophy can increase knowledge about our current world. Initial proposals can be sent to the Editor-in-Chief, Prof. Lorenzo Magnani, lmagnani@unipv.it:

- A short synopsis of the work or the introduction chapter
- The proposed Table of Contents
- The CV of the lead author(s)

For more information, please contact the Editor-in-Chief at lmagnani@unipv.it. Indexed by SCOPUS, ISI and Springerlink. The books of the series are submitted for indexing to Web of Science.

More information about this series at http://www.springer.com/series/10087

Arturo Carsetti

Metabiology

Non-standard Models, General Semantics and Natural Evolution

Springer

Arturo Carsetti
La Nuova Critica
V. Lariana 7, Rome
Italy

ISSN 2192-6255 ISSN 2192-6263 (electronic)
Studies in Applied Philosophy, Epistemology and Rational Ethics
ISBN 978-3-030-32720-0 ISBN 978-3-030-32718-7 (eBook)
https://doi.org/10.1007/978-3-030-32718-7

This Springer imprint is published by the registered company Springer Nature Switzerland AG
The registered company address is: Gewerbestrasse 11, 6330 Cham, Switzerland

To Gregory Chaitin, a great scholar and a gentle teacher

Acknowledgements

This volume could not have taken place without a considerable amount of help. My gratitude goes first to Werner and Elizabeth Leinfellner who died in recent years and whose teaching is still very much alive in me. I am also very grateful to Lorenzo Magnani for agreeing to include the volume in the prestigious series "SAPERE" of which he is the Editor. I am equally grateful to Leontina Di Cecco who examined and approved the manuscript at the editorial level with rare skill and professionalism. I have also been helped enormously by conversations with Franz Wuketits who died recently, Gregory Chaitin, Stephen Grossberg, Henri Atlan, Julian Nida Ruemelin, Giuseppe Longo, Johann Goetschl, Felix Costa, Jean-Paul Delahaye, Dirk van Dalen, Jean Petitot and the late Barry Cooper and Francisco Varela. In particular, I have greatly benefited from discussions with them on the occasion of some specific International Colloquia of *La Nuova Critica* that once again turned out to be the conceptual "skeleton" of the volume. I would like to mention, in particular, the teaching of the friend and Maestro Gaetano Kanizsa which is present albeit subtly in so many parts of the volume.

I am indebted to my collaborators Andrea Cataldi and Pia't Lam for their help at the editorial level. I would like, in particular, to thank Anita Weston for her invaluable advice and suggestions which contributed to the quality of the presentation of the book. Special thanks to Ms. Priyadharshini Subramani, a project manager at Scientific Publishing Services of Springer and her team for their excellent work in handling the editorial production of the book.

Finally, I would like to recall that the beginning of this book came to take shape starting from some notes I wrote in 1970 when Vittorio Somenzi asked me to present to his students at the University of Rome "La Sapienza" the book just released by Monod: *Chance and Necessity*. Once again, I can only express my gratitude to the great (and humble) Master.

Contents

Chapter 1
On the Verge of Life: Looking for a New Scientific Paradigm

Abstract From a general point of view, at the level of life we are constantly faced with an inner self-modulation of a very peculiar coupled system: the alternative splicing represents only one of the essential tools relative to the self-organization of the channel. In this sense, the dialectics between coder and ruler really plays a fundamental role: it is in view of a continuous synthesis of new proteins and new structures as well as of the accomplishment, each time, of a correct assimilation process. In any case, the problem is to establish, every time, a correct relationship between invariance and morphogenesis. Life is hiding in the creative equilibrium at stake, a balance that must be constantly renewed on pain of dissipating life itself. Meaning and incompressibility are the two actors who tread the scene of life determining each time the due balance in accordance with an evolution which concerns the surfacing of the natural world but in the conditions of a continuous (and "intended") metamorphosis.

1.1 Meaning and Incompressibility in the Evolutionary Landscape

When we look at Nature as a craftsman, i.e. as an inventor who works under conditions of constant tinkering and, in general, as morphogenesis in action, we see that the molecular systems that rearrange DNA normally process molecules in accordance with the presentation of a grid of signals. As is well known, retroviral recombination is one of the genetic mechanisms that utilize such grids; at this level we can inspect, in particular, the synthesis of a single DNA molecule from two or more molecules of viral RNA. This task is carried out by an enzyme, the retroviral reverse transcriptase which is "guided" by a multiplicity of signal grids in the original RNA. As the transcriptase synthesizes the replica from its template it may fall under the action of particular grids that trigger specific template switches that present themselves as the key event of retroviral recombination. If we generalize these results taking into consideration more complex techniques as, for instance, DNA shuffling, we can go more deeply into the secret reasons of this kind of recombination. As is well known, those stretches of DNA that do code for amino acids in the proteins are called exons. Exon shuffling is a process in which a new exon is inserted into an existing gene or an

A. Carsetti, *Metabiology*, Studies in Applied Philosophy, Epistemology and Rational Ethics 50, https://doi.org/10.1007/978-3-030-32718-7_1

exon is duplicated in the same gene [1]. It characterizes eukaryotic evolution and has contributed substantially to the complexity of the proteome. Exon shuffling creates new combinations of exons by intronic recombination which is the recombination between two non-homologous sequences or between short homologous sequences that induce genomic rearrangements. The mechanism includes the insertion of introns at positions that correspond to the boundaries of a protein domain, tandem duplication resulting from recombination in the inserted introns, and the transfer of introns to a different, non-homologous gene by intronic recombination. The diversity that can be generated by duplicated exons is strikingly shown by the Down syndrome cell adhesion molecule (*Dscam*) gene in *D. melanogaster*. The multiple, mutually exclusive exons of *Dscam* lead to enormous numbers of splice variants. The secret of this astonishing combinatorial power lies exactly in the intended individuation each time of the adequate grid.

Gene regulation was first studied most accurately in relatively simple bacterial systems. Most bacterial RNA transcripts are said to be colinear, with DNA directly encoding them. In other words, there is a one-to-one correspondence of bases between the gene and the mRNA transcribed from the gene (excepting 5′ and 3′ noncoding regions). However, in 1977, several groups of researchers identified a series of RNA molecules that they termed "mosaics," each of which contained sequences from noncontiguous sites in the viral genome [2]. These mosaics were found late in viral infection. Studies of early infection revealed long primary RNA transcripts that contained all of the sequences from the late RNAs, as well as other specific sequences: the introns, i.e. those stretches of DNA, which get transcribed into RNA but not translated into proteins. The human genome is estimated to contain some 180,000 exons. With a current estimate of 21,000 genes, the average exon content of our genes is about 9. In general, introns tend to be much longer than exons. An average eukaryotic exon is only 140 nucleotides long, but one human intron stretches for 480,000 nucleotides. Removal of the introns—and splicing the exons together—are among the essential steps in synthesizing mRNA. Early in the course of splicing research, yet another surprising discovery was made; specifically, researchers noticed that not only was pre-mRNA punctuated by introns that needed to be excised, but also that alternative patterns of splicing within a single pre-mRNA molecule could yield different functional mRNAs [3]. The first example of alternative splicing was defined in the adenovirus in 1977 when it was possible to demonstrate that one pre-mRNA molecule could be spliced at different junctions to result in a variety of mature mRNA molecules, each containing different combinations of exons. An example of a gene with an impressive number of alternative splicing patterns is the *Dscam* gene from *Drosophila*, which is involved in guiding embryonic nerves to their targets during formation of the fly's nervous system. Examination of the *Dscam* sequence reveals such a large number of introns that differential splicing could, in theory, create a staggering 38,000 different mRNAs. As Schmucker et al. showed in 2000, this ability to create so many mRNAs may provide the diversity necessary for forming a complex structure such as the nervous system [4]. In general, the existence of multiple mRNA transcripts within single genes may account for the complexity of some organisms, such as humans, that have relatively few genes. Alternative splicing exactly provides a mechanism for producing a wide variety of proteins from a small number of genes. While we humans may turn out to have only some 21 thousand genes, we probably make at least 10 times that number of different proteins. It is now estimated that 92–94% of our genes produce pre-mRNAs that are alternatively-spliced. What is more, in the case, for instance, of the *Dscam* proteins, we can observe that they are used to establish a unique identity for each neuron. Each developing neuron synthesizes a dozen or so *Dscam* mRNAs out of the thousands of possibilities. Which ones are selected may appear to be simply a matter of chance, but because of the great number of possibilities, each neuron will most likely end up with a unique set of a dozen or so *Dscam* proteins. As each developing neuron in the central nervous system sprouts dendrites and an axon, these express its unique collection of *Dscam* proteins. If the various extensions of a single neuron should meet each other in the tangled web that is the hallmark of nervous tissue, they are repelled. In this way, thousands of different neurons can coexist in intimate contact without the danger of non-functional contacts between the various extensions of the same neuron [5].

At a basic level, regulation includes splice-site recognition by the spliceosome, which is mediated by various proteins. Additional regulatory levels include environmental changes that affect splice-site choice and the relationship among transcription by RNA polymerase II (RNAPII), nucleosome occupancy and splicing. The existence of introns and differential splicing helps to explain how new genes are synthesized during evolution. As a matter of fact, splicing makes genes more "modular," allowing new combinations of exons to be arranged during evolution. Furthermore, new exons can be inserted into old introns, creating new proteins without disrupting the function of the old gene. In this sense, the definition of alternative exons is very important for understanding the links between splicing and evolution. In accordance with these results, the genome appears definable, in theoretical terms, as a "model", a model finally inducing the full articulation of a "ruler" working as a recipe (as an articulated series of recipes). In the light of this broad definition which will be later clarified, let us assume, from a speculative point of view, that the scanning, the self-identification and the methodical control as performed by the ongoing creative process at the level of the ruler on the basis of the integrated activity of the different protein circuits and the involved regulatory procedures, can lead (according to a metamorphosis that is consumed in the sky of an abstract programming) at the level of the computing (and self-organizing) global protein membrane, to the identification by selection (and intended elimination) of a specific score, i.e. of a coherent and minimal set of instructional constraints (which recalls, for instance, although only in some respects, what happens in the case of percolation phenomena) able to trigger the emergence from the "stone" of new isles of creativity through embodiment. In order to understand the meaning of this last statement that appears dark in itself, let us refer to an ancient Myth, the Myth concerning Marsyas and Apollo, i.e. the relationship between simulation and creation. According to the Myth, Marsyas must offer himself as a stool-grid for the recovery by the God (by the creative process in action) of himself (of his own origins). Marsyas gives the possibility to the God to sink the knife into his depth (his very bark) thus opening to new embodiment through metamorphosis (but on the basis of the score played by Pan). It is only through the sacrifice of Marsyas that the ruler can come to express, in its own metamorphosis, new and unheard isles of creativity, thus allowing the recovery to be carried out.

This is precisely the task of the Silenus, he who leads to the new irruption through both the extroversion of himself and the incisions made by the God at the level of his cortex. It is by means of this engraving that the God can come to manifest himself in his depth, allowing the arising of new incarnation. Only the score engraved on the very skin of the Silenus can lead to a new modulation of the creative song of the God (to the score as played by the Angel in a celebrated painting by Saraceni: "The Vision of St. Francis"). Without the spinet and the simulation work fielded by Marsyas the construction by the God of the lyre as an instrument of creation of natural beings could not have taken place: the God who plays his instrument does not merely simulate or evoke, he actually creates. Here is a creative activity that preludes to (and coincides with) the new embodiment: *Deus sive Natura*. Only if I offer the proper simulation and come to be engraved at the level of my own cortex can I give myself as a grating (cf. "The martyrdom of St. Lawrence" by Titian) for the triumph of divinity. I come to deploy myself in the ongoing simulation of the worlds and I come to be selected, finally arriving at the point of death but having the opportunity to offer the grid able to support a new expression of the Bios. In this way, the Work leads the God in the outcropping of himself, thus giving rise to new Nature. Here is the flourishing of Nature, a flourishing necessarily marked by a Work of art. The God that emerges manages to see but according to the deployment of a possible infinity of vision acts as performed by the different autonomous agents at work (the arising observers). Without the sacrifice of Marsyas the God could not come to regain his own creativity. But it is only when this happens that Marsyas will come to be added to himself by the God as a creator (but in his own same death).

The simulation work put in place is a function of both the selective action taken by the God who proceeds to incise the Silenus' bark, thus succeeding in illuminating his own profound recesses, and the close confrontation performed by the membrane with what constitutes the informational bombardment coming from the external environment. Such complex interplay is in view of enabling the right regulation, i.e. the regulation, able to nurture the coupled system organically ensuring not only the due improvements but also the necessary creative revolutions, the revolutions, that is to say, able to guarantee both the correct recovery and the emergence from time to time of specific novelties albeit in the framework of safeguarding the development of an ever renewed Self. If it is true, as G. Braque states, that the rule must correct the emotion, it will be able to do so only if it comes to be completely declined as such at the level of the self-organization process at play: only meaning that proves productive can open to the new flesh. The emotion that makes itself as a ruler by metamorphosis opens up to the mirroring. In other words, the function that self-organizes does so, of necessity, within the ambit of a meaning that becomes in turn coder in action.

In accordance with these first hints, the identification of the afore mentioned score presupposes the utilization of specific tools (reflexive, simulative, denotational, compositional etc.) living in a logical space characterized by the presence (and the utilization) of specific second-order methods and non-standard procedures

[6]. The mathematics of living beings necessarily articulates (at least) in the context of second-order Cybernetics. By exploiting these procedures at the level of the analysis of the ongoing coupled integration process, it is then possible to individuate specific (and hierarchical) schemes of programs connecting a variety of regulatory factors able to trigger new lines of creative luxuriance at the level of the original source. In other words, these schemes are utilized by the developmental source in order to modulate (and construct) new patterns of expression of its deep information content at the surface level. In this way the action expressed by the polymerases comes to be affected (and determined) from time to time by a variety of primary and secondary "contractions". These patterns appear indissolubly linked to the new game that is opening between exons and introns and between regulatory DNA and structural DNA. Here specific protoprograms get into the action at the deep level of the Bios: we can actually identify the ultimate roots relative to the possible unfolding of a new embodiment process as related to the birth of new genes, new proteins and renewed transcription units. Hence the necessity of recognizing at the level of DNA transcription and related procedures, the existence of general meta-schemes in action concerning the progressive unfolding of both a true editing process and a precise self-modulation activity. The work performed by means of the alternative splicing constitutes one of the essential threads of this very activity. On the basis of our metaphor, we could say that it is actually Marsyas as craftsman of the ongoing simulation who modulates the inner "sense" of the *Via*, helping thus in following this very sense along its secret paths: he works for his "resurrection", but for the Other and in the Other, also exploiting the story-telling of the resurrection itself along the sweep of his artwork. On the one hand, we have a biological (and computational) emerging system that evolves and self-organizes by means of the realization of specific computations and assimilations as performed at the level of its nucleus, while, on the other hand, we have the individuation-invention of intended schemes of programs, the individuation, in particular, of specific grids relative to different transcription complexes each with a different blend of proteins. It is the order of the binding proteins involved in these grids (or signal arrays) that determines and primes each time (in accordance with the action expressed by the different polymerases) the synthesis of new proteins when they are considered as necessary by the coupled system, on the basis of the estimates made, in order to assure its biological equilibrium as well as to explore its potentialities in view of giving the correct answers to the challenges imposed by the environment and guaranteeing the renewal of its creativity. In accordance once again with our metaphor (and with the ancient tale), it is only through the incisions operated on his skin that Marsyas can come to listen to the song of the God. Along the assimilation process the membrane dreams of possible worlds while the God provides to incise it. In general, the incisions are only possible because the membrane is able to read and simulate itself acting as a filter able to "extrude" itself. In other words, it is only the correct exploration of the second-order realm that can allow the true hearing of the song. I do not send only messages signaling the need for the production of proteins already put into yard, I also have to establish the right "suggestions" for the production, along the assembly line and in accordance with the analysis performed at the second-order level, of new proteins on the basis of an

alternative splicing of the exons. Hence the importance of the editing work as well as of the selective sieve. It is only with reference to such a dialectics that the new coder can, each time, take his moves in a creative but intentional way. The new eyes that come then to open, the new arising life of the flesh, will appear, therefore, as related to the measurements put in place by the craftsman by means of the artificial simulation tools (and the ongoing narration) in order to perform an adequate recovery process on the basis of the outlining of the correct gridiron. In this way, the coder will then be able to present itself to itself up to fixate and identify itself as an observing system in action. It appears able to hear the song of the God as this very song throbs of burning flesh of Marsyas. Here is a painter who deploys (and narrates) his very metamorphosis into a renewed Nymph but only in the presence of the melody played by Pan (cf. P. Picasso, "The Pan flute", Musée Picasso, Paris). In the end we shall be faced with the emergence of a body (our body) but we shall also be faced with the deploying of a consciousness, the consciousness concerning the idea on behalf of our mind of our body.

Always in accordance with the ancient Myth, it is the membrane that promotes (performs) the extroversion and that is engraved by the knife acting in ambient creativity which alone can lead (even if indirectly) to unravel and select the plot of the deep creativity modules in action at the DNA level. This kind of performance is in view of identifying both new modulations with respect to the alternative splicing (and the correlated rising of new proteins) and the relaxation of a new relationship between the nucleus and the membrane (as well as between incompressibility and meaning). The Goddess who recognizes herself in the emerging body of Nature (cf. Tintoretto, "St. Mary Magdalene", La Scuola Grande di S. Rocco, Venise) and realizes in herself the right reading of the great stone book written in mathematical characters (which is at the origin of the aforementioned emergence) can come to provide (through appropriate elaborations and simulations) further regulatory patterns able to trigger new forms of creativity capable of giving birth, in turn, to new assemblies of proteins. Here is the legacy of Marsyas at work, his own coming to act as a trigger on the basis of the simulation put in place and here is the sudden emergence of a new song at the level of life. If at the level of the membrane and the software of meaning in action, i.e. at the level of the path of abstraction, one does not reach the correct extroversion in accordance with the Method offered by the God no inheritance will then be possible and no related trigger. There will be no renewal of the roots of life nor the opening of the potentialities inherent in them on the basis of the "intelligent" regulation offered by the membrane. In other words, there will be no possibility of secondary contractions according to a "project" and a renewed integration.

From a general point of view, at the level of life we are constantly faced with an inner self-modulation of a very peculiar coupled system: the alternative splicing represents only one of the essential tools relative to the self-organization of the channel. In this sense, as we have shown in many publications, the dialectics between coder and ruler really plays a fundamental role: it is in view of a continuous synthesis of new proteins and new structures as well as of the accomplishment, each time, of a correct assimilation process. In any case, the problem is to establish, every time, a correct relationship between invariance and morphogenesis. Life is hiding in the creative equilibrium at stake, a balance that must be constantly renewed on pain of dissipating life itself. Meaning and incompressibility are the two actors who tread the scene of life determining each time the due balance in accordance with an evolution which concerns the surfacing of the natural world but in the conditions of a continuous (and "intended") metamorphosis.

1.2 DNA Regulation and Control: The Interface Between Ruler and Coder

At the biological (and creative) level, the original, developmental and self-organizing source, while transmits and applies its message, constructs its own structure. The transmission content is represented by the progressive "revelation" through forms of the very source, of the self-organizing (and emotional) "instructions" concerning, each time, its actual realization at the surface level and its primary operational closure. This closure realizes its own in variance and, at the same time, its metamorphosis by means of the full unfolding of a specific embodiment process, by means of replication and by means of a continuous production of varied complexity. In particular, the source extends continuously itself as supporting hardware also on the basis of a continuous incorporation activity. The final result of this process cannot be seen as a simple output: "the phenome" (according to the terminology proposed by Atlan [7]) completely determined by an input string (the genome). It by no means represents, however, an output determined only by a mapping function, so that the resulting structure appears as not encoded at all in the input string. On the contrary, the transmission of the information content on behalf of the source appears to be a process of realization, revelation and "renewal" of itself, a process realized, in particular, in accordance with the conditions proper to a coupled system: as a matter of fact, at the biological level the function necessarily self-organizes together with its meaning thereby avoiding (as linked to the stake) Scylla (the simple dissipation) and Carybdis (the pure crystallization). Hence the necessity at the level of life of the continuously renewed realization of a specific compromise: the "aperiodic crystal", as Schrödinger called that particular intermediate state represented by DNA. Within this theoretical framework, the action of the *telos* is principally instrumental in enabling a progressive renewal of original creativity; the victory over the forces of dissipation primarily depends on the capacity on behalf of the source to manage to realize a targeted (and constrained) development, at the formal level, of its own potentialities (according to the conditions of the experience): the *telos* constitutes the method and the *Via* of this particular unfolding thus animating the deep patterns of the embodiment process. The need for meaning selection to intervene derives from this, enacting a Medusa-like use of the reflexivity tools to fix the lines of a realm of necessity able to guarantee the achievement and the permanence of an ever-new (and fruitful) equilibrium. Ariadne prepares the kingdom of the "ruler" by helping in offering adequate orderings and semantic constraints, by fixing and petrifying, and thus preparing the necessary support for the new "conception". Through successful embodiment the model therefore realizes within itself (but according to a surfacing process) the deep content of the original incompressibility while opening it to the dominion of meaning and "rational perception".

This realization represents, in this sense, a process of liberation from the phantoms of repetition, from the closure in oneself, from one's own abandonment to the impossible dream of a total self-reflection (Narcissus). It is in this sense that, as advocated by H. Bergson (quoted by J. Monod in *Chance and Necessity*), God needs men to affirm his creativity in a constantly renewed way.

Actually, the DNA appears as the receptacle of information "programmed" by natural selection. It is the engine of evolution, the true engine, in particular, of the embodiment process, along the successive expression of the laws of the "inscription" and in view of the full unfolding of the realm of necessity, the actual triumph of the "ruler". It articulates through the cellular growth that is taking place according to the constraints imposed by the selection performed in ambient meaning and by the *bricolage* operated with respect to the pre-existing structures. It is along this peculiar channel that the flux of deep creativity may, therefore, express itself and tune, in an innovative way, its original incompressibility according to the outcrop at surface of different stages of functional construction and to the correlated realization of a specific "nesting" process at the semantic level. Within the frame of this particular kind of metamorphosis the genome must be seen as neither a program nor a set of "data". It appears, on the contrary, definable, in theoretical terms, as a "model" as we have just said, a model finally inducing the full articulation of a "ruler" working as a recipe (as an articulated series of recipes). Both the interpretation function and the representation apparatus concerning that particular cellular machinery represented by the activity of proteins make essential reference to this kind of model. We are effectively in front of a complex cellular (and living) "network" within which we can individuate, first of all, the presence of a specific process of "inscription" as well as of an interpretation function operating at the level of surface representation. This network is intrinsically open to the fluxes of deep creativity and results constrained by the selective pressures acting in ambient meaning. The role of the attractors takes place in the background of this intricate series of processes; it cannot only concern a component of the cycle of the metamorphosis. The genome expressing itself gives rise to a surface structure, to the body of that particular factory represented by the living cell of which the genome itself comes to inhabit the nucleus (at the level of eukaryotic cells) acting as the template and the forge of the transformation. The original meaning that manages to express itself can only be incarnated in correspondence to specific cognitive schemes.

The genome expresses itself into a given phenotype in a complex way. Actually, at the basic level, the genome sequence codes for its own translating machinery. It determines the birth of a cellular machinery responsible, in turn, for gene regulation and expression. A particular gene, for instance, codes for RNA polymerase whose function is to transcribe the genes into messenger RNA. Without RNA polymerase there is no messenger RNA, we are faced with the absence of cellular life. However, RNA polymerase is necessary for its very synthesis because it transcribes its gene. Hence the essential circularity that characterizes living organisms [8]. The cellular machinery "represents", step by step, the genome into an organism realizing the final stage of what we have called the embodiment process. In this sense, the genome and the cellular machinery really interact by establishing an evolving and coupled network: as we shall see one of the key results of this interaction is represented by the continuous engraving (through selection) at the level of the organisms of specific formats: among them we can distinguish, first of all, the formats relative to the architectures of sensorial perception. As Bray correctly remarks: "In unicellular organisms, protein-based circuits act in place of a nervous system to control behaviour; in the larger and more complicated cells of plants and animals, many thousands of proteins functionally connected to each other carry information from the plasma membrane to the genome. The imprint of the environment on the concentration and activity of many thousands of proteins in a cell is in effect a memory trace, like a 'random access memory' containing ever-changing information about the cell's surroundings. Because of their high degree of interconnection, systems of interacting proteins act as neural networks trained by evolution to respond appropriately to patterns of extracellular stimuli. The 'wiring' of these networks depends on diffusion-limited encounters between molecules, and for this and other reasons they have unique features not found in conventional computer-based neural networks" [9]. According to Bray, a common feature of protein circuits in living cells is represented by their ability to integrate multiple inputs. We can find the most astonishing evidence of this combinatorial capacity at the level of the regulation of DNA transcription in eukaryotic cells. Actually, a typical gene in a multicellular organism requires the assembly of a transcriptional complex composed of enzymes, transcription factors and gene regulatory proteins. Because these components are drawn from a very large pool of candidates, an extremely large variety of different transcriptional complexes, each with a different 'blend' of proteins, is possible.

As we have just said, transcription is carried out by an enzyme called RNA polymerase and a number of accessory proteins called transcription factors. Transcription factors are proteins involved in the process of converting, or transcribing, DNA into RNA. Transcription factors include a wide number of proteins, excluding RNA polymerase, that initiate and regulate the transcription of genes. Transcription factors can bind to specific DNA sequences called enhancer and promoter sequences in order to recruit RNA polymerase to an appropriate transcription site. Together, the transcription factors and RNA polymerase form a complex called the transcription initiation complex. This complex initiates transcription, and the RNA polymerase begins mRNA synthesis by matching complementary bases to the original DNA strand. The mRNA molecule is elongated and, once the strand is completely synthesized, transcription is terminated. The newly formed mRNA copies of the gene then serve as blueprints for protein synthesis during the process of translation. Regulation of transcription is the most common form of gene control. The action of transcription factors allows for unique expression of each gene in different cell types and during development. All species require a mechanism by which transcription can be regulated in order to achieve spatial and temporal changes in gene expression. It is not easy to understand the relationship between the genome and the functioning of cells: scientists have primarily devoted their attention to the study of genome products, namely proteins and expressed RNAs, such as tRNA and mRNA. Proteomics is the study of the set of proteins in a cell or tissue, and it includes details on protein quantity and diversity. However, the proteome may not tell a cell's entire story. Actually, proteins are dynamic and interacting molecules, and their changeability can make proteomic snapshots difficult to understand. Furthermore, there are many technical challenges in characterizing molecules that cannot be easily amplified and have several post-translational modifications. Thankfully, taking into account the intermediate step between genes and proteins, i.e. the transcripts of messenger RNA, bridges the gap between the genetic code and the functional molecules that run cells [10].

The cells of higher organisms exhibit an incredible number of genetic responses to their environment. This is largely the result of TFs (transcription factors) that govern the way genes are transcribed and RNA polymerase II is recruited. In addition, by working in combination with chromatin, TF signals can exert a finer level of control over DNA by allowing for gradations of expression. TF families further increase the level of genetic complexity in eukaryotes, and many TFs within the same family often work together to affect transcription of a single gene. Given the function of TFs, along with other mechanisms of eukaryotic gene regulation, it is not surprising that complex organisms are capable of doing so much with so few genes. It is these processes, more than the number of genes, that come to differentiate complex and simple organisms from each other from a genetic standpoint. From a general point of view, transcription factors bind in a combinatorial fashion to specify the on-and-off states of genes; the ensemble of these binding events forms a regulatory network, constituting the wiring diagram for a cell: transcription factors thus appear as proteins that control the production of other proteins. Moreover, enzymes involved in the same pathway in the cell are often controlled by the same transcription factor. In this sense, in order to account for the complexity of the cell, we have, *in primis*, to remember that the phenotypic behaviour of a cell depends on how those proteins that result from gene expression go on to interact with one another. Protein interrelationships drive the structure and function of cells, including how cells react to changes in temperature, nutrients, and stress [11].

In multicellular organisms, nearly every cell contains the same genome and thus the same genes. However, not every gene is transcriptionally active in every cell—in other words, different cells show different patterns of gene expression. These variations underlie the wide range of physical, biochemical, and developmental differences seen among various cells and tissues. Thus, by collecting and comparing transcriptomes of different types of cells or tissues, we can gain a deeper understanding of what constitutes a specific cell type.

A transcriptome represents that small percentage of the genetic code that is transcribed into RNA molecules—estimated to be less than 5% of the genome in humans (cf. Frith et al. [12]). The proportion of transcribed sequences that are non-protein-coding appears to be greater in more complex organisms. In addition, each gene may produce more than one variant of mRNA because of alternative splicing, RNA editing, or alternative transcription initiation and termination sites. In this sense, the transcriptome captures a level of complexity that the simple genome sequence does not. By studying transcriptomes, it is possible to generate a comprehensive picture of what genes are active at various stages of development. Complexity of transcriptional control can be well illustrated by comparing the number and locations of cis-control elements in higher and lower eukaryotes. For instance, *Drosophila* typically has several enhancers for a single gene of 2–3 kb, scattered over a large (10 kb) region of DNA, while yeast have no enhancers but instead use one UAS sequence per gene, located upstream. Long-range regulation is thought to be indicative of the need for a higher level of control over genes involved in cell development and differentiation. The yeast genome encodes around 300 TFs, or one per every 20 genes, while humans express approximately 3000 TFs, or one per every 10 genes. With combinatorial control, the twofold increase in TFs per gene actually translates into many more possible combinations of interactions, allowing for the dramatic increase in diversity among organisms. TFs are not solely responsible for gene regulation; eukaryotes also rely on cell signaling, RNA splicing, siRNA control mechanisms, and chromatin modifications. However, TFs that bind to cis-regulator DNA sequences are responsible for either positively or negatively influencing the transcription of specific genes, essentially determining whether a particular gene will be turned "on" or "off" in an organism. By means of expressing particular genes and proteins, cells differentiate into various mature cell types, whether sensory neurons, muscle cells, or red blood cells and so on. In this sense, the regulation of genes via transcription factors and chromatin is as important as the presence and nature of the genes themselves. Promoters are the sequences of DNA that determine when a gene is expressed. These sections of DNA sit in front of genes and provide a 'landing site' for transcription factors (proteins that switch gene expression on and off) and RNA polymerase (the protein that reads DNA and makes an mRNA copy). Different promoter sequences have different strengths, and genes with 'strong' promoters are expressed at a higher level than those with 'weak' promoters. A single TF can regulate the expression of many different genes. Within this framework, the alternative splicing can give rise to different forms of mRNA. Moreover, under specific conditions the introns are not degraded thus giving rise to other functional RNAs at the level of gene regulation, i.e. non-coding RNA (ncRNA). In this field the research is in progress: what seems clear, in any case, is that the problem inherent in gene regulation turns out to be more and more intriguing and pervasive every day. In particular, it now requires an increasingly massive recourse to methods of analysis and models that stem from the theory of complexity, from the theory of self-organization and from various sectors of non-standard mathematics, a recourse-appeal, in particular, that can only call into play specific functions such as those relating to the observation and control on the part of the self-organizing system of what is its own growth path as it articulates on the basis of the *bricolage* at work. The *telos* becomes part of the dynamics put in place and requires special analytical tools to be handled with due care.

With respect to this frame of reference as well as, for instance, to the last results obtained by the ENCODE scientists as regards the regulatory action put in place at the level of DNA, let us underline once again that, from a general point of view, the transcription factors bind in a combinatorial fashion to specify the on-and-off of genes and that the ensemble of these very factors forms a regulatory network constituting the wiring diagram of the cell. In this way, a limited set of transcription factors is able to organize the large diversity of gene-expression patterns in different cell types and conditions. The co-association of the different transcription factors is highly context-specific and can be organized in accordance with a stratified (but dynamic) hierarchy. It results in a grid capable of inducing each time a specific set of secondary "contractions" at the transcription level. The transcription factors orchestrate gene activity from moment to moment according to specific basic rules on the basis of the application of particular self-organizing modules. The first map of the regulatory protein docking sites on the human genome as obtained in the frame of ENCODE Project reveals the dictionary of DNA words and permits a first sketch of the genome's programming language. There is an enormous variety of features that contribute to gene regulation. Every cell type uses different combination and permutation of these features to generate and "individuate" in a self-organizing way its unique biology. In this context, we must, as humans, not only individuate the secret reason of the orchestration put in place, but also to participate in it: we have to be actors and spectators at the same time as advocated by Prigogine [13].

The genome determining the expression of a cellular machinery, determines the birth of both an apparatus and a surface program embedded in that apparatus. Effectively, the apparatus doesn't appear to be an interpreter with a given program, it appears rather as a parallel computing network (but working at the surface level) with a precise evolving internal dynamics, a network able, moreover, to represent and reflect itself (and express, still within itself, its own truth predicate). The program "embedded" in this apparatus concerns the general frame of the connections and constraints progressively arising, its exclusive capacity to express (and canalize by attractors) a specific coordination activity within the boundaries of the becoming net. This capacity, on the other hand, can be "crystallized" on the basis of specific operations of self-representation and abstraction, so that it can be, finally, seen as the very "image" of the functional (but embodied) synthetic modules in action through which the apparatus progressively self-organizes thus expressing its autonomy. Here we can find the real roots of an effective identification process through adequate "enlightening". At this process-level we can find, on the one hand, information which codes for proteins, and, on the other hand, an operational closure which leads to the individuation of a "self"; from this derives the effective possibility of a dominion of necessary abstraction scanned by the "time" of the embodiment and supported by the cellular body in action. According to the ancient Myth, while Narcissus goes beyond the mirror and petrifies arriving to inhabit the waters of meaning and acting as a fixed point, Marsyas, instead, burns itself giving rise to productive abstraction. Narcissus, in turn, offering the right attractors (along the first sense of the *Via*) allows the rising of that holistic enlightening that preludes to the new conception, i.e. to that particular breath of meaning that leads to the renewed birth of Marsyas (taking on the guise of the Lord of the garlands).

The legacy of Marsyas acts as a trigger: the original "memory" burns through a host of intensities, it comes therefore to be channeled according to imagination. The DNA comes to emerge in a functional sense when through the trigger operated by Marsyas (namely through the score played by Pan), the hardware related to the original "memory" comes, finally, to articulate in accordance to the deployment of an autonomous activity of imagination: memory + imagination. When this happens (but necessarily in the presence of adequate orderings) here is the giving of the risen: the memory self-organizes together with its language. Marsyas causes the memory as supporting hardware to articulate according to imagination acts based on that particular primer represented by his own spoils: hence the emergence of a new ruler. Here is the possible rise of a DNA that reveals itself as the true support of the way to the incarnation. There is a precise circularity: without DNA there is no real life and real embodiment, changing, however, the conditions of the channel will change the inspiration concerning life itself. Language that becomes thought in life in relation to the surfacing of creativity, and thought that becomes vision in the truth, at the level of the full nesting of meaning. The artificial that is made natural comes, therefore, to recover itself but in the affirmation of its own creativity: meaning may recover completely itself only becoming productive. It is at that moment that the pure simulation language as it had been expressing through Marsyas' inventions (following the performed extroversion and the *experimentum* put in place) finally comes to act as a trigger. We assist to an embodiment process, to an artificial thought that comes to enjoy again the roots of life in the flare up of the flames (the *Sylva*). New intensities will come to arise at the biological level finally articulating, under the aegis of imagination at work, as specific (conceptual) functions of synthesis. We will have the giving of information of a productive (and semantic) nature which will come, then, to express completely itself at the level of the full construction of a natural body. The source that comes to articulate in ambient meaning can only open up to the development of a mind and the activity of mind's eye. Here is a function that comes to self-organize together with its meaning. Marsyas as coder and pure software of the simulation comes to turn (as a result of the metamorphosis so well described at the level of the ancient Myth) into the new Nymph which manages to get out of the spoils of the Silenus. Hence the emergence of a very special library of "functions", a library of functional modules that act, in particular, at the biological level coding, in particular, for proteins by means of transcription, translation, etc. A library which appears, as we have just seen, immersed in an essential, ineliminable circularity; it is in the framework of this circularity that the phenomena characterizing life and cognition are given. Such circularity cannot be circumscribed in terms of a simple program in itself isolated, however complex it may be. The incompressibility that constitutes the landscape in which it expresses itself is constantly growing on itself: the relationship between Nature and Work (between natural and artificial) plays, at this level, a fundamental role. There is no absolute Chaos followed, therefore, by a Reflexivity also having an absolute character. Only when we reveal ourselves as prisoners of a given conception of Reflexivity do we end up seeing a specific incompressibility as absolute.

1.3 Simulation Procedures and the Evolution of the Inner Membrane Complex

As Bray states, there exists a precise role of the membranes in guiding the opening of the genome and in coming to modulate the "viscera". But in view of achieving this, we need a map, a patient work of "embroidery" (as suggested by Vermeer in his paintings) that can only go through the extroversion of what is the operative articulation of the membrane itself. Marsyas must realize the extroversion of the operative articulation of his cortex in view of assuring the God of a correct and renewed modulation of his creativity: a growth in the recovery. It is necessary to open the viscera according to cell growth but in harmony with the environment. This is in view not only of the invariance but also of a correct morphogenesis. In particular, Marsyas has to work for the correct recovery in order to come to be added as a creator. Only, in fact, such an addition can ensure the non-self-abandonment by the God, his coming out of the blood labyrinth. There is no assembly for the assembly but rather the intelligent montage depending on the correct growth (in the recovery) of the original creativity.

The membrane is to be locked in a functional sense only if such growth is ensured. It will therefore be necessary to refer to complex computations, self-organization motifs, second-order methods, etc. You will need to be able to offer yourself as a stool in order to carry out the adequate montage on your skin but this, in turn, implies the need to climb on your shoulders. Marsyas must confront the God through the stool (cf. Balthus "*Grande Composition au corbeau*"). The God does not have to mirror himself but to recover himself: he must operate the successful montage with reference to the models offered by Marsyas. The track (and the trigger) is given by the flute sound or better by its overcoming from within: the operated simulation opens the Way to creative modulation which can only prevail if it can prove able to open to a renewed Nature without any arrest at the only mannerist exaltation of the Work.

In this way Marsyas must be subjected to selection by the God in view of being added as a creator. Hence the irruption and the rising of the new Nymph (cf. Titian, "Nymph and Shepherd", Kunsthistorisches Museum, Vienna). Here is the one who turns his gaze to the Painter-Midas on the verge of dying and who proves able to explore both the path of incarnation and that of observation. Here the alternative splicing is revealed as a function of the possible construction of new arrangements in view of what will then be the delineation at the level of complex biological organisms of a mind, of an action aimed at driving the detachment in view of the achievement of a renewed autonomy. It is the membrane that gradually contributes to the constitution of the mind. Hence the unique development of a cell by means of a set intertwined of self-organization processes, a cell that "knows" and that expresses itself intentionally. Here's a piece of information no longer anchored to Information Theory as baptized by Shannon and possessing a definite intentional character. In this sense, as we shall see, circularity gives rise to the completion of the detachment and to the closure in the distinction. It is looking at the ancient remains of the Silenus that the Minotaur manages to act as an observer recognizing himself in the canvas but in the Other. The new coder prepares the ground for the journey of the Minotaur and the emergence of his eye, the eye of his mind. It is in ambient Reflexivity that the outcrop of an I is given along its drowning in the image. The Minotaur opens his eyes and discovers his mortality: *Et in Arcadia ego*. On the other hand, it is with reference to the identification of the eigenvalues on the carpet that Marsyas comes to extrude his cortex coming to be engraved by God.

In this sense, it is always with reference to a process of metamorphosis and to the correlated embedded "program" that the genome can act as a model. A model that must not be considered only from a logical and semantical point of view (in a denotational sense), but also from a biological and functional point of view. As a model, that is, considered as instructional information + intentionality, finally inducing the emergence of the ruler in action. In order to describe the functional nature of this particular model as well as of the link existing at the biological level between form, information and enformation, the resolution, however, of at least of four orders of problems results indispensable: (1) the outlining of a statistical mechanics at the biological level concerning genes and macromolecules and no more only atoms and molecules, able, moreover, to take into consideration the role of the self-organization forces at play; (2) the outlining of a semantic information theory taking into consideration the concept of observational meaning: the meaning as connected, at the same time, to a process, to an observer and to a hierarchical and embedded representation process; (3) the outlining of new measures with respect to the very concept of biological information. We need measures capable of taking into the consideration the growth processes, the statistical fluctuations living at the microscopic level etc. The Shannonian measure concerns essentially stationary processes articulating in a one-dimensional landscape, on the contrary, a true measure of information for life and hereditary structures should concern semantic information at work as connected to the action of specific, self-organizing coupled systems in continuous evolution; (4) the identification in terms of the theory of complex systems and of the second-order Cybernetics of the role of control and modulation performed by the membranes at the cellular and intercellular levels.

The model is the "temporary" receptacle of the biological functions and the replicative life; in particular, it appears, as we have just said, as the receptacle of an information programmed by natural selection. The genome, in other words, is a model for a series of biological actions and symmetry breakings, for the realization of a complex path whose goal is represented by the attainment, on behalf of the apparatus, of a sufficiently complete functional autonomy at the surface level (within a dynamic ambient meaning). The interpretation function relative to this kind of model appears exactly to concern the actual realization of the embodiment process through the progressive execution of meaning. The realm of the bodies in action is the realm of a ruler able to settle the contours of meaningful information. In this sense, as Maynard Smith correctly remarks [14], a DNA molecule has a particular sequence because it specifies a particular protein: it contains information concerning proteins and specifies a form that articulates as synthesis in action. DNA and (regulatory) proteins carry instructions for the development of the organism; in particular genomic information is meaningful in that it generates an organism able to survive in the environment in which selection has acted. In turn, the organisms act as vehicles capable of permitting the source the successive realization of its own "renewal". The source "channels" itself through the *telos* finally articulating itself as a model: we are faced with intentional information at work, really immersed in its meaning. In this way the ruler is continuously connected to the action of the coder as well as to the progressive articulation of a specific nesting process at the semantic level. Everything is marked by a complex regulatory interplay involving a continuous and circular link between depth information and surface information and between meaning and incompressibility.

The coupling constituted by the apparatus and the surface "program" possesses, as we have just remarked specific capacities of emergent self-representation and articulated invention; it paves, in particular, the way to the unfolding of a possible simulation activity. This coupling realizes itself in a rehearsal stage in strict connection with the model and interacting with it. If the coupling succeeds (under the guidance expressed by the *telos*) in realizing adequate simulation programs able to "open" and support the successive expression of protoprograms living at the level of deep creativity (if, that is, the teleonomic "dream" of the model comes true), it will grow and gain a form of functional autonomy. So the ancient model will become mere data and a new functional order will irrupt (exactly at the moment when the apparatus starts to behave as a pure interpreter, processing DNA as data with reference to a surface program which results nearly completely "exteriorized"). Life is a self-organizing process supported by the continuous transaction of two original forces: incompressibility and meaning. It is at this level that we are necessarily faced with the riddle concerning morphogenesis. In this context the simulation work, if successful, constitutes each time the yeast that presides over the emergence of the new coder.

The coder imparting intentionality allows information to be articulated as semantic, to be immersed in meaning, i.e., to sanction the birth of an apparatus able to perceive according to the truth. The progressive realization of the embodiment, the very insurgence of an apparatus able to feed meaning, corresponds to the coding in action. But the original source will manage to code because the *telos* was able to "follow" (and execute) meaning in an adequate way. Insofar as the DNA constitutes itself as a model *via* the embodiment process, the model at work necessarily reveals itself as intentional (self-organizing, in perspective, as a possible biological basis of a specific perceptual activity). Hence a source that through the *Via* manages to code and perceive but with respect to the progressive articulation and the "adjunction" of specific "observers" that continuously join (and inhabit) *Natura naturata*. Then, the rising of a new "conception" will finally be possible at the level of the effective closure of operant meaning as well as the birth of specific simulation procedures articulating as a whole on the basis of the development of body's "intelligence". The source that posits itself as model renders itself to life but in accordance with the truth. Only the *telos* capable of reflecting itself into the truth will be able to offer the source real intentionality: hence the intrinsic circularity between form and information.

From an objective point of view, the inscription process possesses a self-limiting character with respect to the infinite potentialities of expression which are present in the original source. Moreover, the model, at the beginning, is "blind" (from the point of view of categorial intuition) like the Minotaur as envisaged by Picasso in Vollard Suite. The Minotaur through the embodiment must be filled by intuitions and semantic constraints, only in this way he can progressively manage to open his eyes. In order to become a suitable channel for the successive revelation of the deep information living in the source, the model must not simply replicate itself: it also has to prepare and utilize specific generative tools in order to enable the source to express its inner creativity in a new and more complex way. It necessarily self-organizes within an ambient meaning on the basis of *telos*' activity but in order to permit the renewal and the recovery of the very tools of original incompressibility. The original source can transmit and apply itself (the instructions concerning its own construction), only to the extent that it succeeds in canalizing, each time, by exploiting first of all the "inscription" process, that peculiar emergence of a new type of semantic order which selectively inscribes itself by programs into the becoming fibers of the channel and which shows itself as a tool for a new "moment" and a new expression of the deep creativity proper to the very source. The crucial step, along this difficult path, is represented by the actual development (through the complete realization of the embodiment and at the level of a non-replicative realm) of a specific process of recovery. The essential aim of this kind of process consists in extracting and outlining, according to different stages of exactness and completeness, a simulation in formal terms (a sort of simulation machine) of the functional procedures which constitute from within that particular biological network that finally presents itself as a concrete achievement of the embodiment process (and as an observing system [15]). The opening of the eyes at the level of the Minotaur coincides with his petrifying and fixing as an observing system and with his becoming a mortal being, a body of computations and assimilations. In contrast, the simulation implemented by Marsyas leads to the building of an artwork (on a conceptual level) that allows the entry into the realm of abstraction as a prodrome for the renewal of his very being. The simulation activity is first of all in view of a correct assimilation of the external cues: we assimilate in order to give the correct answers and to develop our potentialities also modifying our very roots Narcissus has to fix his image by invariants in order that meaning can arrange itself in accordance with a holistic enlightening finally accepting him in its interior as dreaming Endymion. Hence the possibility of the occurrence of a new conception, the possible birth of a simulation process capable of ensuring a self-renewal of the system but in the Other, and in the overcoming of itself, an overcoming which finally coincides with the ultimate expression of the ruler.

The biological advantage represented by the capacity of simulation developed on behalf of the nervous system, along the course of evolution, is well known. The existence of this kind of mechanism transcends the genetic scheme of Darwinian selection. Normally the simulation activity is considered as the only available instrument in order to make adequate predictions about the behavior of non-linear dynamical systems. In this way the organism can possibly avoid difficulties, conflicts etc. and maintain its stability. The problematics, however, is more subtle. The evolving (and living) net, actually, simulates itself (through emergent self-representation processes) in order both to think of itself "from the outside" and to offer, finally, the abstract programs emerging from the simulation machine to the deep generative fluxes as a *recipe* for a new, coherent and productive revelation of their informational content. So the simulation activity appears biologically programmed (for what concerns its primary aim) not only to control and prevent disturbing events possibly occurring in an external reality, but also to offer a possible "guide" to the processes of emergent (and intentional) information in order to avoid forms of incoherence, collapses from the inside etc. and to express new patterns of action consistently. Thus, we have to distinguish once again between a selection performed in ambient incompressibility and a selection performed in ambient meaning. It is this last kind of selection that supports the new irruption, the irruption of a new "order" (of the invariants characterizing the emerging chaos) at the level of the realized metamorphosis: hence the new coder in action. In this sense, it is the simulation work that permits the assimilation of external cues. Narcissus has to fix himself in the image by invariants in order to prime the nesting process relative to meaning in action, hence the possibility of the rising of a new conception on the basis of the offer on behalf of Narcissus of the adequate cipher concerning the eigenforms.

In this sense, for instance, the simulation machine, considered as a schematic assembly of simulation programs identified also by means of specific assimilation procedures, can allow, on the basis of the utilization of reflexive and non-standard tools, virtual elements and relations living at the level of the original source to be individuated, so that it will be possible to test, at a selective level, new forms of composition and synthesis of these same elements and relations (but along their effective emergence). If this attempt results successful, the ancient model will become (in the limit) pure data and, at the same time, new forms of inscription will graft themselves on its connective tissue. Then, new generative fluxes and new (correlated) emergent properties will appear, new patterns of coherence will rise and, contemporarily, possible conflicts between the invariance characterizing the ancient inscription and the emergent expression of new generative principles will be canalized according to renewed and evolving correlation functions. The simulation activity appears devoted, first of all, not to guarantee a given fixed invariance, but to prepare the emergence of new and more complex forms of invariance at the creative (and chaotic) level: thus the semantic source at the level of the ruler appears to construct itself, step by step, as a "productive" channel. It self-regulates and finally appears not only as a ruler in action, but also as a co-evolving channel, as a reality indissolubly linked to a specific process of self-generation of meaning in accordance with the truth. In this sense, the original biological source attains its own invariance not because it reflects a given, fixed order (an order that, in the background of the dissipation process, could only present itself as the order or law of Chance), but because it succeeds in developing, each time, the necessary tools for its representation at the surface level so that new levels of deep incompressibility can, finally, express themselves by means of the emergence of new functional (and living) "ideas". These ideas will represent the "intentional" stakes able to canalize the real development of the capacity of productive regeneration of the source, the new "moments" of a Time considered, contemporarily, both as construction and as recovery. They reveal themselves completely only at the moment of the opening-up of the new coder and represent the true results of the application of specific conceptual procedures. They flourish as sensitivity in action. Thus life and cognition appear as indissolubly intertwined.

Invariance is the fruit of the move to the standard, but in view of the new irruption. The difficulty lies in recovering each time the correct path in accordance with the dialectics in action: Hercules at the crossroads of dissipation and crystallization. I must inscribe myself (to tap my flesh, as in the ancient tale) because the Other could succeed in writing and dictating to me in the interior of my soul and simultaneously to "ascend": hence the need for a new concept of information as productive complexity. What changes now is the relationship with the incompressibility: the information is no longer a given, with regard to which a measure can be defined; no longer the result of a hinge-action defined within itself. The actions of measuring, observing and inventing enter the realm of incompressibility, determining its internal evolution according to the unfolding of an increasingly diverse dialectics among the original selective forces and between invariance and morphogenesis. This explains why the link between depth information and surface information has to be re-identified each time, and why the range of the instruments of the theory of incompressibility has to be extended in a continuous way. In this sense, as we have just seen, at the level of a biological network acting as a ruler, it is necessary to postulate the existence, not only of pathways of observation, but also of simulation, in view of the possible opening up of a new language. Hence the necessity of the recourse to the mathematics of non-standard: life is within life and precedes it, but in proportion to the correct construction of the Artwork, and the successive edification of a Nature which can then open to new patterns of perceptual activity.

In the frame of this new theoretical perspective we need more adequate measures of meaningful complexity, capable, for instance, of also taking into account the dynamic and interactive aspects of depth information. We have, in particular, to outline new models for the interface existing between the observer and the observed system. At the level of these kinds of models, emergence (in a co-evolutionary landscape) and truth (in an intensional setting), for many aspects, will necessarily coincide. Moreover, a coupled system in this theoretical perspective must be considered as a multiplicative unit: a source-forge for other coupled systems. With respect to these types of systems we have, first of all, to recognize the presence of a computational apparatus able, in particular, to operate on symbols and schemes. Actually, we well know, for instance, how it is possible to define a particular O-machine, the halting function machine, that has as its "classical" part the universal Turing machine specified by Turing in 1936 and as its "non classical" part a primitive operation that returns the values of Turing's halting function $H(x, y)$. The halting function machine can compute many functions that are not Turing computable. In particular all functions reducible to the halting function and/or the primitives of an ordinary Turing machine are computable by the halting function machine.

If we take into consideration an O-machine we can easily realize that each additional primitive operation of an O-machine is an "operation on symbols" and we know that the notion of a symbol-processing machine is more general than the notion of Turing machine. In this sense, we can think of the functions that are solvable by a first-order oracle machine as being harder that those solvable by Turing machine. From a more general point of view, oracles should be considered as information with possible error that take time to consult [16]: a physical experiment directly presents itself as an external device to the Turing machine. We are faced with a first individuation of a coupled system, as originally envisaged by Turing. Besides the distinction of degree, we also have to underlie that if it is true that self-reflection (as well as limitation procedures) is part of biological functioning and that self-reflection inscribes itself in the very structure of the different undecidables, it is also true that we can distinguish, at the level of the afore-mentioned computational apparatus, many other "intellectual" instruments. Among these instruments we can individuate, for instance, the coagulum functions, the simulation recipes, specific operations on schemes etc.

By exploring the territories of simulation, the system opens to the paths of morphogenesis: on the one hand, we have invariance but with respect to meaning in action, on the other hand, we are faced with morphogenesis but in accordance to a continuous recovery of the roots proper to the original creativity. In this sense, from a biological point of view, the genome, no more considered only as a pure replication system but also as a self-organizing system devoted to the expression and the "renewal" of the character of creative generation proper to the source, appears to articulate its construction activity according to precise and differentiated stages of development. Its production of varied complexity (as it is obtained on the basis of the progressive unfolding of the coder) is moulded by specific selective pressures according to particular creative arrangements of nets of constraints: finally, for its part, the ruler can determine, in ambient meaning, (under *telos*' guidance) the progressive expression of a specific channel; it paves the way for a further partial revelation of new forms of incompressibility. It self-organizes as a ruler acting, at the same time, as a recipe. The new order possibly arising in consequence of this very revelation will inscribe itself within and over the old structure, within and over the biochemical "fibers" of the last revealed order. In this sense, we can remark how it is impossible, in principle, according to these considerations, to recreate present life by simulation. Man can only make himself as a tool in order to determine, utilizing suitable simulation models, a further and different expression of the Bios.

The depth information (incompressibility) of which we speak about is not a particular structure or an order or a set of correlations. It is the "place" (the locus of action) of the regulative principles, of the *ideas* that arise in the background of the irruption of new chaos. These regulative invariants become an order only when they succeed in expressing themselves as a specific act of synthesis within the frame represented by the net of constraints determined by the progressive realization of an adequate self-organizing channel. In other words, this type of synthesis can take place only with reference, first of all, to the dialectics in action between the production of varied complexity, on the one hand, and the progressive nesting of meaning's selective activity, on the other hand. The actual manifestation of this dialectics, of this interactive dynamics, shows itself as a process of production of meaningful information, a process which necessarily articulates through the successive appearance of specific patterns of order and coherence as well as of specific at tractors. We have, moreover, to underline, once more, that it is, precisely, the autonomous capacity, at the level of the channel of articulating according to these patterns in a correct way (but exploiting, step by step, the suitable paths and "reflecting" the general scheme of these very paths in order to build an adequate simulation program), that can, finally, open the way to the revelation of unexpressed proto-programs living at the level of the deep generative fluxes as well as to the subsequent transformation of the generative principles into specific properties able to generate the information intrinsic to the dynamical systems. As we have just said, only in this way can the coder offer to the biological source real intentionality.

1.4 The Unraveling of Meaning in the Sky of Abstraction

It is the successful Work that anticipates, therefore, the flourishing of new Nature, the new expression of creativity and the new outcrop. Without recourse to the revisable thought and the simulation activities there may not be new observation in accordance with the truth: *Verum* and *Factum*. Truth is to the extent of the Work (and of the Time) and vice versa. The proteome must simulate in itself, on a diagrammatic level (the level of its conceptual complexity), the genome: i.e., Marsyas must simulate *Natura Naturans* until he manages to appear superior to it in terms of the modulation performed, but precisely because of this he will prove able to allow the God to come to create in a renewed way in the overcoming of the ongoing simulation and in adherence to the accuracy of the incisions made. Here is the ultimate meaning of the stool's offer as expressed by Balthus in the painting: "*Grande Composition au corbeau*" already mentioned: no stool, no re-establishment of creativity and no break-in could be possible just as without the cipher no reflection and no conception could come to take place.

It is the God who prepares for that new categorial through which Marsyas resurrecting to life as a new Nymph will look at his ancient remains proving, finally, able to bypass them at the level of the ongoing renewed incarnation: Marsyas revealing himself capable of climbing on his shoulders will thus prove able to work on his metamorphosis at the level of the whale's belly. This transmutation of Marsyas gives rise to the birth of new observation (the Nymph) but on the basis of the trigger at work as it comes to rise on the basis of the same death of the Silenus. The genome programs for proteins but in the presence of continuous firing changes in the network: here is the coming to light of a targeted modulation activity that presides in perspective to the construction of new proteins but in function of new creativity and new Nature (in the overcoming of Marsyas). The Painter-membrane must be able to simulate himself in order to offer the stool, he must, in particular, proffer himself as a grid in the extroversion: only if this happens new Nature can come to be born. Each time new invariance must be found, but in view of a new opening to abstraction: in S. Lawrence by Titian the stool is a grid of tortured flesh. This is the offer that allows the irruption, i.e. the coming out by the God from the abandonment of himself and the very beginning of the activity of detachment. Otherwise the invariance can only result in simple repetitiveness. Hence the centrality of the issue relative to the observer: this is why the Painter must paint his own image as an observer, but on the basis of the primeval work performed by Marsyas (thus remaining always at the level of the revisable thought). The only way for the Painter to go out to meet reality is to die in the Other but in view of his resurrection.

By painting Marsyas anticipates, he thinks about the possible arising of new observation finally activating it by working for the involved transmutation through his coming to climb on his own shoulders. Then comes the moment when he will come to be overwhelmed: it will be, then, the new Nymph who will come to observe him. Insertion of exons and creative modulation of DNA by means of the alternative splicing: it is necessary to let the viscera speak, but the right ones, those that allow the insertion of the contraction modules according to a fruitful growth in creativity. Hence the selection and the necessary descent into the Underworld. Marsyas can only act as a creator if he manages to simulate well. It is only in this way that he can come to be added in his own transmutation. As added, he will turn out to be part of an ongoing creation, but at that moment he can only transmute into the Minotaur along the realization of the metamorphosis and of the successive detachment: as Picasso shows, man constitutes, in turn, the ultimate effect of the ongoing detachment of the Minotaur. In this sense the Minotaur is the true alter-ego of Picasso (as Painter).

The genome that self-organizes in the context of the regulatory activity as expressed by the membrane finally leads to the surfacing of deep biological information through the "intended" construction of specific assemblies of proteins. It is precisely in reference to this outcrop that the possibility emerges for a Minotaur to come to constitute himself as biological autonomy. Here is a Minotaur who comes both to self-regulate himself by means of awareness in action and to govern his self-reflection on the basis of the emergence of a mind, i.e. of a "faculty" of imagination articulated by eigenforms. This fact, in turn, will allow the Minotaur to perform the "detachment" in accordance with both self-distinction and the link postulated by H. von Foerster: "I am the observed relation between myself and observing myself" [16]. At the level of the Metamorphoses as outlined by Picasso the eigenforms emerge, in particular, as "digital" images and in them Narcissus–Minotaur recognizes himself drowning in the waters thus reaching his invariance through the fixing of an ideal identity that is summarized in a cipher, i.e. in the flower that will be offered to the Goddess in view of new conception.

On the contrary, when we are faced with the nesting process, the membrane appears involved in itself (in its firing diagram) up to develop not a mind acting as the ordering of the ongoing constructions but a simulation activity according to the techniques of the variation as elaborated at the level of the cortex (cf. Proust, *La Recherche*). We are faced with the arising of a possible myriad of different I as well as of a multiplicity of possible worlds considered as different variants that run after each other. Hence the necessity no longer of the order imposed by an analytic "I think" but of the progressive individuation of a Self understood as the work of sewing at the deep level of the arising different I in view of the conquest by the hero of an ideal nucleus, of that coherent modulation, that is to say, of his inner structures (as it manages to emerge to the extent of the nesting at work) that reveals itself as essentially related to the continuous coming to climb on the part of the hero on his shoulders. Through the revisable thought in action (Marsyas) the embroidery of the different I (of the divided I) gives rise to the Self of a soul. Hence a Work that unfolds as a cathedral, a Work that lurks and nests in the secrets of the soul and the air (cf. De Nittis) determining the rise of new Nature.

Here is the lacemaker of Vermeer-Dalì, she who along her seam work stands (and acts) as an incessant factor of unitary growth with respect to her cortex, thus leading back to a Self but with a thousand implications: in this way we can take a first look at what is the skeleton underlying the development of natural intelligence: a kind of embroidery but realized in the depths of a soul in the presence of two sources of illumination (as well illustrated by Vermeer): an inside source and an external source. The first one concerning the canalization of life, the second one Grace in action. If Marsyas is the hero who builds the Work coming to act as the engine and the plot of the ongoing nesting, as the one, that is to say, who modulates this same nesting with his pipe, this is, however, possible because the hero proves able to preside over the extroversion of himself with respect to his membrane-bark, that particular extroversion that will enable the continuous recourse by the hero to this same bark (but parcelled out) in accordance to a simulation plan (to a series of anticipations and resurrections which could lead to the formation of a thought in life.). The lacemaker shows precisely how self-organization at the cognitive level can only succeed through the patient and enlightened pursuit of the unraveling of meaning.

It will then be possible to enter the process of outcropping itself on the basis of precise hints, this fact may also lead, where necessary, to the construction of new proteins but making incessant recourse to secondary contractions, which, however, appear to be linked to a continuously revisable design (to a grammar of the abstraction: the dialectic as outlined by P. Veronese). Hence the journey in the Underworld as well as the necessity to preside over a continuous reset in the edges of chaos with possible birth of new triggers following the different incinerations of Marsyas: here is the legacy of the simulation and the real possibility of a new primer each time. The God will then add to himself Marsyas as creator: he will be able to go back to the roots of his creativity and at the same time to see-observe himself at the level of his progressive surfacing (to the extent of the coming up of the Nymph). On the one hand, a Deity creates the Silenus in order to add to itself a new creator (a possible multiplicity of creators), on the other hand, a Deity irrupts so that in the depths of the *Sylva* a new Minotaur can come to stand with a view to add himself as an observer to the Deity. Hence a Deity crossed by a multiplicity of perceptual acts. The Goddess can observe to the extent that Narcissus adds himself to the Temple. The God can create to the extent that Marsyas comes to be added to the burning bush.

It is Marsyas-Painter who dies that allows all this guiding, through the trigger, the self-organization process proper to the Nymph. Here is the hero who identifies himself in the rising of the Nymph but at the level of the praxis of art (and not coming out of it if not in the successive real transmutation). Hence the effective emergence of the God as well as the path pursued by the mind of the Minotaur with a view to the new birth of Marsyas on the basis of the reflection by the Goddess in the body of Nature. Picasso who paints his incarnation actually "finds-meets" Marie Therese as his bride in the rediscovered flesh. He comes to be observed by the woman thus preparing, by means of this very observation, what will be his metamorphosis. Working for mirroring he overshadows the new conception.

As the ancient Marsyas the Painter will have to come to die, but his death must be consumed first of all at the level of the praxis of art. It will then be necessary for a new Goddess and a new God to come into the field. New observation skills will come to be born from within. We pass through multiple observations-resurrections that are necessary for the perpetuation of life. This means, once again, that the DNA changes but on the basis of the guide offered by Pan: Pan guides the transformation of the original memory back to new DNA. By changing the articulated provisions of the mRNA and the RNA on the basis of a simulation plan we could find ourselves in front of a completely renewed DNA as a result of the self-organization process at work: at this level a new observation activity could come to peel away from the simulation in progress.

With respect to this frame of reference, Marsyas reveals himself as meaning in action that comes to act as thought in life. The "conversion" will be given for various levels (see "St. Paul Conversion" by Tintoretto), but the coming up of the new observation will always be at the level of the Other: the hero's passing is always in the Other and for the Other: here is an Ideal that turns into the Real. From Merleau-Ponty to Varela: the project and its link with Autopoiesis. It is the Work that anticipates Nature but when the Way realizes itself in the truth, the *Factum* proper to the observation can only be revealed in itself as true reality (acting as such at the level of the community of the living beings). We will therefore have a Nature that unfolds from itself through the articulation of the Work: here is a Nature that grows on itself but in view of new abstraction. We will come, in this sense, to be inhabited (along our very metamorphosis) by new meanings and new genes according to a continuous *bricolage* that is at the foundation of the relationship between Incompressibility and Meaning. Here are lives that are given to the biological level but also to the artistic and social level. Here is a transformation that requires continuous work of embroidery.

References

1. Gerstein, M. B., et al. (2012). Architecture of the human regulatory network derived from ENCODE data. *Nature, 489*, 91–100.
2. Berget, S. M., et al. (1977). Spliced segments at the 5′ adenovirus 2 late mRNA. *Proceedings of National Academy of Sciences, 74*, 3171–3175.
3. Darnell, J.E., Jr. (1978). Implication of RNA-RNA splicing in evolution of eukaryotic cells. *Science, 202*, 1257–1260.
4. Schmuker, D., et al. (2000). *Drosophila Dscam* is an axon guidance receptor exhibiting extraordinary molecular diversity. *Cell, 101*, 671–684.
5. Keren, H., Lev-Maor, G., & Ast, G. (2010). Alternative splicing and evolution: Diversification, exon definition and function. *Nature Reviews. Genetics, 1*, 345–355.
6. Carsetti, A. (2012). *Eigenforms, natural computing and morphogenesis*. Paper presented to the Turing Centenary Conference, CiE 2012.

7. Atlan, H. (2000). Self-organizing networks: Weak, strong and intentional, the role of their under determination. In A. Carsetti (Ed.), *Functional models of cognition* (pp. 127–143). Dordrecht: Kluwer A.P.
8. Kourilsky, P. (1987). *Les artisans de l'hérédité*. Paris: O. Jacob.
9. Bray, D. (1995). Protein molecules as computational elements in living cells. *Nature, 376*, [p. 309].
10. Gerstein, M. B., et al. (2012). Architecture of the human regulatory network derived from ENCODE data. *Nature, 489*, 91–100. Sakabe, N. U., & Nobrega, M. A. (2013). Beyond the ENCODE project. *Philosophical Transactions of the Royal Society B, 368*, 20130022.
11. Mayran, A., & Drouin, J. (2018). Pioneer transcription factors shape the epigenetic landscape. *The Journal of Biological Chemistry, 293*, 13795–13804. https://doi.org/10.1074/jbc.r117.001232.
12. Frith, M., Pheasant, M., & Mattick J. S. (2005). The amazing complexity of the human transcriptome. *European Journal of Human Genetics, 13*, 894–897.
13. Prigogine, I. (1980). *From being to becoming*. San Francisco.
14. Maynard Smith, J. (2000). The concept of information in Biology. *Philosophy of Science, 67*, 177–194.
15. Carsetti, A. (2000). Randomness, information and meaningful complexity: Some remarks about the emergence of biological structures. *La Nuova Critica, 36*, 47–109; Carsetti, A. (2013). *Epistemic complexity and knowledge construction*. New York: Springer; Carsetti, A. (1987). Teoria algoritmica della informazione e sistemi biologici. *La Nuova Critica, 3–4*, 37–66.
16. von Foerster, H. (1981). Objects: Tokens for (eigen-) behaviors. Observing systems, the systems, inquiry series. Salinas, CA: Intersystems Publications, [p. 279].

Chapter 2
Drawing a Software Space for Natural Evolution

Abstract The genome expresses itself into a given phenotype in a complex way. Actually, at the basic level, the genome sequence codes for its own translating machinery. It determines the birth of a cellular machinery responsible, in turn, for gene regulation and expression. A particular gene, for instance, codes for RNA polymerase whose function is to transcribe the genes into messenger RNA. Without RNA polymerase there is no messenger RNA, we are faced with the absence of cellular life. However, RNA polymerase is necessary for its very synthesis because it transcribes its gene. Hence the essential circularity that characterizes living organisms. The cellular machinery "represents", step by step, the genome into an organism realizing the final stage of what we call the embodiment process. In this sense, the genome and the cellular machinery really interact by establishing an evolving and coupled network: as we shall see one of the key results of this interaction is represented by the continuous engraving (through selection) at the level of the organisms of specific formats: among them we can distinguish, first of all, the formats relative to the architectures of sensorial perception. In this way the circularity proper to life necessarily unfolds in the conditions of Reflexivity.

2.1 Autopoiesis and Self-organization. The Operational Closure of Nervous System

Starting from the revolution brought on by J. Monod at the level of molecular Biology also by means of the graft he operated on the body of this field of research of the methods and instruments deriving from classical Cybernetics and Shannon's Information Theory, the great issues related to circularity and feed-back procedures as they are given at the level of the Bios, have come to acquire, as we have seen, an ever greater importance, thus opening up to the construction of renewed models with a view to achieving a deeper understanding of the mechanisms of life.

From Monod we passed on to Jacob and the interaction between the possible and the actual to finally arrive, at the end of a wild cavalcade that touched on all topics related to contemporary molecular Biology, to the great issues related to the alternative splicing, the ENCODE Project, the synthetic life as outlined by Craig

A. Carsetti, *Metabiology*, Studies in Applied Philosophy, Epistemology and Rational Ethics 50, https://doi.org/10.1007/978-3-030-32718-7_2

Venter, the CRISP methodology etc. It is necessary, however, to point out that some of the issues that characterized the above-mentioned cavalcade on the more strictly biological and experimental level, such as, for instance, the issues concerning the circularity of life, the process of Self-organization and so on, were already, in the early seventies, to the attention of many scholars who—in comparison to the biologists more directly engaged on the experimental front—worked as scientists "next door" proving to be, in any case, able to make further and important contributions, above all at the theoretical level, to the development of molecular Biology. These contributions have come to open new horizons with respect to the traditional formulations of the concepts of information and complexity thus allowing to investigate more deeply the self-organization phenomena characterizing the development of the Bios. In particular, on the basis of these contributions it was, therefore, possible to correlate the model initially introduced by Monod with the methods and the principles proper to the second-order Cybernetics. Among these strands, an important role is certainly still played by the doctrine of Autopoiesis as it has been outlined since the seventies by H. Maturana and F. Varela.

According to Maturana and Varela circularity is at the base of the biological processes. Their attempt to frame this circularity on a theoretical level is based to a large extent on the adoption of specific quantitative and informational tools. In particular, their doctrine has enlightened some of the pregnant aspects of this circularity both in relation to the construction of the Self and the outlining of the membrane. What is, however, even more important is the fact that the two scholars came to give a first mathematical characterization of the aforementioned circularity by recourse to the mathematics of self-reference and self-organizing circuits; hence the engagement of their doctrine in the wider fields of Reflexive domains, Denotational Semantics, second-order Cybernetics etc. Starting from the original papers published by Maturana and Varela, the seventies and the eighties saw the blossoming of a series of very important researches ranging from Maturana and F. Varela to S. Kauffman, H. von Foerster, L. Kauffman, A. Carsetti etc. In this context, the path of research pursued by F. Varela gradually took on an increasingly central role, as rich as it was of fruitful suggestions. The early eighties mark, in particular, a turning point quite important to what concerns such a path, a path that had seen the great scholar primarily engaged in the previous decade in the definition of the mathematical foundations of the doctrine of Autopoiesis. This turning point is consumed in large part when Varela is moving to Europe in order to continue and deepen his research. It is in the eighties, in fact, that some important papers by Varela are published just as Varela flies several times from the Americas to France and Italy as visiting professor. In 1984, in particular, a special issue of the Journal *La Nuova Critica* (the Italian Journal for the Philosophy of Science whose Scientific Committee he was called to take part at the invitation of V. Tonini) entitled: "*Autopoiesi e teoria dei sistemi viventi*" was published in Italy in the framework of his work as visiting professor at the University of Rome and under his direct supervision [1]. This issue of *La Nuova Critica* opens with an article by Varela expressly translated by him from French into English on the occasion of the publication of this special issue of the Journal. At the heart of this choice, there was the will both to submit a paper that appeared to be a turning point in the scientific life of Varela, to the attention of a wider audience of scholars (thus providing, in particular, an adequate summary of the versatility of the research carried out by the great scholar in the years that preceded his move to Europe) and to insert it in the context of a collection of papers devoted to the doctrine of Autopoiesis also in view of a critical comparison with the theoretical approach provided at the computational and mathematical level by Denotational semantics.

For Varela, the issue represented the way both to be confronted with the Denotational Semantics as outlined by D. Scott (and to which a specific attention had been dedicated by A. Carsetti in the article entitled: "*Semantica denotazionale e sistemi autopietici*" included in the aforementioned issue of *La Nuova Critica* on the basis of what has been agreed by the Editor of the Journal with Varela) and to develop his research in the field of the mathematics of Self-reference but in unitary connection with Brouwer's fixed point theorem.

As is well known, the scientific vision of Varela was profoundly influenced by the analysis of the volume by Spencer Brown concerning the Laws of the Form published in 1969 [2]. It is precisely the common interest in this volume which had, on the other hand, constituted the trait-d'union that initially had come to tie the Chilean scholar to the great American mathematician Louis Kauffman. As Louis Kauffman remarks: "Prior to that I had written a paper using Francisco's "Calculus for Self-Reference" to analyze the temporal behavior of self-referential circuits [L. H. Kauffman (1978)—Network Synthesis]. My papers were inspired by Varela's use of the reentering mark in his analysis of the completeness of the calculus for self-reference that he associated with that symbol. I also started corresponding with Francisco, telling him all sorts of ideas and recreations related to self-reference. We agreed to meet, and I visited him in Boulder, Colorado in 1977. There we made a plan for a paper using the waveform arithmetic. This became the paper "Form Dynamics", eventually published in the Journal for Social and Biological Structures [L. H. Kauffman and F. Varela (1980)]. Francisco based a chapter of his book "Principles of Biological Autonomy" on form dynamics. I remember being surprised to find some of my words and phrases in the pages of his book. The point about Form Dynamics was to extend the notion of autonomy inherent in a timeless representation of the reentering mark to a larger context that includes temporality and the way that time can be implicit in a spatial or symbolic form. *Thus the reentering mark itself is beyond duality, but implicate within it are all sorts and forms of duality from the duality of space and time to the duality of temporal forms shifted in time from one another, to the duality of form and nothingness "itself"*. I believe that both Francisco and I felt that in developing Form Dynamics we had reached a balance in relation to these dualities that was quite fruitful, creative and meditative. It was a wonderful aesthetic excursion into basic science. This work relates at an abstract level with the notions of autonomy and autopoiesis inherent in the earlier work of Maturana, Uribe and Varela [H. Maturana, R. Uribe and F. J. Varela (1974)]. There they gave a generalized definition of life (autopoiesis) and showed how a self-distinguishing system could arise from a substrate of "chemical" interaction rules. I am sure that the relationship between the concept of the reentering mark and the details of this earlier model was instrumental in getting Francisco to think deeply about Laws of Form and to focus on the Calculus for Self-Reference.

Later developments in fractal explorations and artificial life and autopoesis enrich the context of Form Dynamics. At the time (around 1980) that Francisco and I discussed Form Dynamics we were concerned with providing a flexible framework within which one could have the "eigenforms" of Heinz von Foerster [Heinz von Foerster (1981)] and also the dynamical evolution of these forms as demanded by biology and by mathematics. It was clear to me that Francisco had a deep intuition about the role of these eigenforms in the organizational structure of biology. This is an intuition that comes forth in his books" [3].

Following his move to France and the work done at the CREA, Varela came to develop a body of work that still leaves us amazed. In particular, the research he carried out in the fields concerning not only the mathematics of self-reference but also the nature of the enactive mind appears very incisive. These areas of research, starting from the first setting offered by the Chilean scholar, came soon to be part of the territories of election of a multiplicity of researchers working in the field of cognitive sciences. The conviction that since the seventies animates the analysis carried forward by Varela in this areas of research is linked, first of all, to the intuition on his part of the indissoluble connection existing between life and cognition, a connection that, in his opinion, is at the basis of the same circularity of life. It is this conviction that binds him deeply to the studies carried out by H. von Foerster and Louis Kauffman and it is precisely this same basic conviction that will lead Varela to confront, in a renewed way, when necessary, the problems related to the relationship between cognition and reality both from a biological point of view and from a cognitive point of view A striking testimony of the importance of Varela's work in this field of research is represented by the fact that Varela's intuitions concerning the link between Enactivism and Realism, (intuitions that were also born on the base of some deep theoretical contributions by Prigogine as well as of some more ancient observations by von Foerster) now come to reveal their convergence, albeit limited and purely speculative, with some of the theses of the so-called participatory Realism as outlined by some brilliant theoretical physicists who work at level of Quantum Mechanics (cf. for instance Christopher Fuchs) but in the wake of some primitive intuitions by B. de Finetti: "*La prévision: ses lois logiques, ses sources subjectives*", *Ann. Inst. H. Poincaré*, 7: 1–68 [4].

To fully understand all the fecundity and the amplitude of the horizons which will gradually be opened at the international level starting from the moment of the first unfolding of Varela's work at the CREA and to come to understand the importance of such work for the development of the studies concerning the individuation of specific mathematical methods able to shed light on the nature of mind, we need to take care of the paper written by Varela in collaboration with Andrade (Cf. J. Andrade, and F. Varela (1984), "Self-reference and fixed points", *Acta Applic. Matem*, [5]). As we have just said, following the studies carried out in America on the basis of the fundamental contributions by Spencer Brown as well as in the footsteps of the primitive intuitions by H. von Foerster (as revisited by L. Kauffman), Varela at the time of his transfer to Europe felt the need to put into safety, so to speak, the work previously carried out at the theoretical level in the field of the mathematics of self-reference and self-organizing circuits. This indeed was the only route available to him with a view to adequately open his mind to the new world represented by Phenomenology. Only once had he reached this goal, did he fell able to deal with the great themes of Reflexivity albeit under the perspective of the embodied cognition and Enactivism. This in turn will open to the possibility of revisiting the original doctrine of Autopoiesis in the light of the studies carried out in the research field of Neurophenomenology. Hence the importance in the eyes of Varela of the paper written with Andrade, a paper intended to insert, as we have already mentioned, the reflexive domains in the unitary frame offered by Brouwer's theorem.

In this way, not only was he able to come into contact even more incisively with Denotational Semantics which had arisen since the seventies at the hands of D. Scott, but also to find, on a strictly formal level, precise cues for what will be, then, the continuation of his research concerning the role played by the recursive processes at the level of the constitution of the eigenforms.

The CREA soon became a popular destination for many scholars from America who felt the need to come to deal critically with the new climate of thought that was coming to birth. First of all D. Dennett, the Dennett who, as a proud champion of reductionism, comes to Paris as visiting professor at the CREA to deepen his research on the nature of mind in accordance with a close confrontation with both Varela and, ideally, M. Merleau Ponty. But next to Dennett and other well-known scholars there were also many young American Ph.D., as, for example, Cristine Skarda, who, on the instructions of their teachers, came to visit the CREA with great joy of Varela and with the birth also of triangulations of studies between La California, Paris, Brussels and Rome. (cf. the issue of *La Nuova Critica. Nuova Serie* N. 18). In this context, the volume by F. Varela, E. Thompson and E. Rosch *The Embodied Mind* (1991, The MIT Press, London), is of paramount importance. The volume was preceded by two major papers: J. Soto-Andrade and F. Varela (1990), "On mental rotations and cortical activity patterns: A linear representation is still wanted", *Biological Cybernetics*, e F. Varela (1990), "Between Turing and quantum mechanics there is a body to be found", *Beh. Brain Sci. (Commentary)*. In these two papers Varela not only explores the mathematics of self-reference but also addresses his attention to the theoretical principles of the second-order Cybernetics in the footsteps of von Foerster [6].

As Varela writes in the Introduction to the volume: "We like to consider our journey in this book as a modern continuation of a program of research founded over a generation ago by the French philosopher, Maurice Merleau-Ponty… We hold with Merleau-Ponty that Western scientific culture requires that we see our bodies both as physical structures and as lived, experiential structures-in short, as both "outer" and "inner", biological and phenomenological. These two sides of embodiment are obviously not opposed. Instead, we continuously circulate back and forth between them. Merleau-Ponty recognized that we cannot understand this circulation without a detailed investigation of its fundamental axis, namely, the embodiment of knowledge, cognition, and experience. For Merleau-Ponty, as for us, embodiment has this double sense: it encompasses both the body as a lived, experiential structure and the body as the context or milieu of cognitive mechanisms" [7].

Varela assumes that the cognitive subject alias the Self is fundamentally fragmented, divided, or non unified It is with reference to such a conception that Varela will then introduce the definition of what is enactive: "In the enactive program, we explicitly call into question the assumption-prevalent throughout cognitive science that cognition consists of the representation of a world that is independent of our perceptual and cognitive capacities by a cognitive system that exists independent of the world. We outline instead a view of cognition as embodied action and so recover the idea of embodiment that we invoked above. We also situate this view of cognition within the context of evolutionary theory by arguing that evolution consists not in optimal adaptation but rather in what we call natural drift. This fourth step in our book may be the most creative contribution we have to offer to contemporary cognitive science" [8].

In this context, Varela is to dive in the words by Merlau Ponty: "When I begin to reflect, my reflection bears upon an unreflective experience, moreover my reflection cannot be unaware of itself as an event, and so it appears to itself in the light of a truly creative act, of a changed structure of consciousness, and yet it has to recognize, as having priority over its own operations, the world which is given to the subject because the subject is given to himself .... Perception is not a science of the world, it is not even an act, a deliberate taking up of a position; it is the background from which all acts stand out, and is presupposed by them: The world is not an object such that I have in my possession the law of its making; it is the natural setting of, and field for, all my thoughts and all my explicit perceptions" [9] As Varela underlines: Mind awakens in a world. We awoke both to ourselves and to the world we inhabit. We come to reflect on that world as we grow and live. According to Merleau Ponty: The subject is inseparable from the world, but from a world which the subject itself projects. Hence the ultimate secret of that essential circularity that permeates the cognitive activity Thus we come to witness the convergence of Varela's main thesis with the new evolutionary view as advocated by D. Lewontin.

In 1993 the *American Journal of Psychology* published a review by Daniel C. Dennett of the volume *The Embodied Mind.* As the great scholar *in primis* remarks: "cognitive science proclaims that in one way or another our minds are computers, and this seems so mechanistic, reductionistic, ... unbiological. It leaves out emotion, or what philosophers call qualia, or value, ... It doesn't explain what minds are so much as attempt to explain minds away". Dennett acknowledges the importance of the enactive proposal and the congruence between this same proposal and the general theses outlined by Lewontin but he cannot hide his puzzlement. "There is something to this, of course—he will write more later—but just how important is it? What are the relative proportions of organismic and extra-organismic contributions to the "enacted" world? It is true, as Lewontin has often pointed out, that the chemical composition of the atmosphere is as much a product of the activity of living organisms as a precondition of their life, but it is also true that it can be safely treated as a constant (an "external", "pregiven" condition), because its changes in response to local organismic activity are usually insignificant as variables in interaction with the variables under scrutiny" [10]. These words well testify as the debate between

Dennett and Varela was not based on prejudices but on the precise reference to experimental data and mathematical proofs.

As we have just said, Varela and Maturana with their doctrine of Autopoiesis had enlightened, from the outset of their analysis, some of the pregnant aspects of that particular circularity characterizing the living beings both in relation to the construction of the Self and the outlining of the membrane. After the publication of the book *The Embodied Mind* in 1991, J. Brockman edited in 1995 a collective volume entitled *The Third Culture* in which are present essays by Varela, Dennett, Kauffman etc. It is precisely in his contribution to the volume that Varela comes to delineate in a perhaps more incisive way both the sense of circularity that is given at the biological level and the sense of that circularity which is given, in its turn, at the cognitive level, thus arriving to identify their indissoluble unity. The creative circle comes to close but at the same time the identity of a new philosophical doctrine achieves its most appropriate image.

As Varela remarks: "Autopoiesis attempts to define the uniqueness of the emergence that produces life in its fundamental cellular form. It's specific to the cellular level. There's a circular or network process that engenders a paradox: a self-organizing network of biochemical reactions produces molecules, which do something specific and unique: they create a boundary, a membrane, which constrains the network that has produced the constituents of the membrane. This is a logical bootstrap, a loop: a network produces entities that create a boundary, which constrains the network that produced the boundary. This bootstrap is precisely what's unique about cells. A self-distinguishing entity exists when the bootstrap is completed. This entity has produced its own boundary. It doesn't require an external agent to notice it, or to say, "I'm here." It is, by itself, a self- distinction. It bootstraps itself out of a soup of chemistry and physics …. In order to deal with the circular nature of the autopoiesis idea, I developed some bits of mathematics of self-reference, in an attempt to make sense out of the bootstrap—the entity that produces its own boundary. The mathematics of self-reference involves creating formalisms to reflect the strange situation in which something produces A, which produces B, which produces A. That was 1974. Today, many colleagues call such ideas part of complexity theory. The more recent wave of work in complexity illuminates my bootstrap idea, in that it's a nice way of talking about this funny, screwy logic where the snake bites its own tail and you can't discern a beginning. Forget the idea of a black box with inputs and outputs. Think in terms of loops. My early work on self-reference and autopoiesis followed from ideas developed by cyberneticists such as Warren McCulloch and Norbert Wiener, who were the first scientists to think in those terms. But early cybernetics is essentially concerned with feedback circuits, and the early cyberneticists fell short of recognizing the importance of circularity in the constitution of an identity. Their loops are still inside an input/output box. In several contemporary complex systems, the inputs and outputs are completely dependent on interactions within the system, and their richness comes from their internal connectedness. Give up the boxes, and work with the entire loopiness of the thing. For instance, it's impossible to build a nervous system that has very clear inputs and outputs" [11]. Hence the almost immediate idea of applying the logic of emergent properties proper to circular structures in order to investigate the functioning of the nervous system. Another scientific revolution enters the scene: a revolution focused on the analysis of the self-organization processes that inhabit the most hidden recesses of brain functioning.

"The consequence is a radical change in the received view of the brain. The nervous system is not an information-processing system, because, by definition, information-processing systems need clear inputs. The nervous system has internal, or operational, closure. The key question is how, on the basis of its ongoing internal dynamics, the brain configures or constitutes relevance from otherwise non meaningful interactions. You can see why I'm not really interested in the classical artificial-intelligence and information-processing metaphors of brain studies. The brain can't be understood as a computer, in any interesting sense, and I part company with the people who think that the brain does rely on symbolic representation. The same intuitions cut across other biological fields. Deconstruct the notion that the brain is processing information and making a representation of the world. Deconstruct the militaristic notion that the immune system is about defense and looking out for invaders. Deconstruct the notion that evolution is about optimizing fitness to live in the conditions present in some kind of niche. I haven't been directly active in this last line of research, but it's of great importance for my argument. Deconstructing adaptation means deconstructing neo-Darwinism. Steve Gould, Stuart Kauffman, and Dick Lewontin, each in his own way, have spelled out this new evolutionary view. Lewontin, in particular, has much appreciated the fact that my work on the nervous system mirrors his work with evolution" [12].

To the volume *The Third Culture* will then follow in 1999 another volume entitled: *Naturalizing Phenomenology. Issues in Contemporary Phenomenology and Cognitive Science*, edited by Jean Petitot, Francisco J. Varela, Bernard Pachoud, and Jean–Michel Roy, a volume that sees the presence of many qualified scholars who had collaborated with the CREA such as Petitot, Roy, Thompson, Petit, Longo, Dupuy etc. The volume represents an important challenge: with its publication it is not only Varela who comes to terms with Husserl's heritage: It is a large share of the scientific French context which comes to confront this philosophical legacy.

2.2 Reflexivity and Imagination at Work: The Role of Eigenforms

Whilst in the last days of his life Varela continued to open up unexplored horizons at the level of an area of research that was increasingly to be characterized in cognitive terms (with special attention to the ongoing discoveries in the field of Neurophenomenology), Louis Kauffman, in the meantime, was developing, throughout his last fruitful twenty years of research, a more and more incisive deepening at the mathematical level of the issues concerning the reflexive domains (and in general the cybernetic study of Reflexivity) in agreement, in particular, to the intuitions introduced in this field of research by H. Foerster by means of the utilization of the methods proper to the second-order Cybernetics.

"If science is to be performed—he writes—in a reflexive domain, one must recognize the actions of the persons in the domain. Persons and their actions are not separate. if an action is a scientific theory about the domain, then this theory becomes a (new) transformation of the domain. In a reflexive domain, theory inevitably affects the ground that it studies. The fact that an entire reflexive domain can be seen as an eigenform suggests the observation of that domain in a wider view. For example, physics can be seen as a reflexive domain and one can take a meta-scientific view, allowing physics itself to be one of the objects of a larger domain in which it (physical science) is one of the eigenforms. Once cybernetics is defined in terms of itself, it becomes what is commonly called "second-order cybernetics." From the point of view of this article, I identify cybernetics as second-order cybernetics and take it (cybernetics) to be itself a reflexive domain. in this way, cybernetics is both the origin of the concept of reflexive domain and is itself an exemplar of that concept in George Spencer Brown's book Laws of Form (Spencer Brown 1969) a very simple mathematical system is constructed on the basis of a single sign, the mark, designated by a circle or a box or a right-angle bracket. i shall not develop the formalism here. We can take < > to stand for the Spencer Brown mark. But that very sign, < >, in our eyes, makes a distinction in the plane in which it is drawn and the interpretation of Spencer Brown's calculus has us understand that the sign refers to the distinction that the sign makes (we make that distinction when we are identified with the sign). Thus, the sign of distinction in the calculus of Spencer Brown is self-referential. The equation < > = < > can be understood as an eigenform equation. We interpret the mark on the left as a transformation from the void of its inside to the marked state that is seen on its outside. We interpret the mark on the right as a mark of distinction. The identity of the mark of distinction with the act of distinguishing is the fundamental eigenform of Laws of Form. it is a conceptual fixed point, not a notational one, and this means that there is no excursion to infinity in this eigenform. The sign indicates the very distinction that the sign makes. Everything is said in the context of an observer. The observer herself makes the distinction that is its own sign in the form, the sign, the first distinction and the observer are identical. The key point about the reflexivity of the sign of distinction in Laws of Form is that it does not lead to an infinite regress. It is through concept that our own thought is kept from spiraling into infinite repetition. Eigenform occurs at the point that infinite repetition is replaced by fixed point and by concept in this way, our perception is always a precise mixture of sense data and the sense of thought" [13]. Hence a deep relationship with the Fixed Point Theorem (that tells that every A in a reflexive domain has an eigenform).

As Louis Kauffman remarks in another paper: "In fact, ultimately, the form of an "object" is the form of the distinction that "it" makes in the space of our perception. In any attempt to speak absolutely about the nature of form we take the form of distinction for the form (paraphrasing Spencer–Brown 1969). It is the form of distinction that remains constant and produces an apparent object for the observer. How can you write an equation for this? The simplest route is to write $O(A) = A$. The object A is a fixed point for the observer O. The object is an eigenform. We must emphasize that this is the most schematically possible description of the condition of the observer in relation to an object A. We only record that the observer as an actor (operator) manages through his acting to leave the (form of) the object unchanged. This can be a recognition of the symmetry of the object but it also can be a description of how the observer, searching for an object, makes that object up (like a good fairy tale) from the very ingredients that are the observer herself. This is the situation that Heinz von Foerster has been most interested in studying. As he puts it, if you give a person an undecideable problem, then the answer that he gives you is a description of himself. And so, by working on hard and undecideable problems we go deeply into the discovery of who we really are. All this is symbolized in the little equation $O(A) = A$. We can start anew from the dictum that the perceiver and the perceived arise together in the condition of observation. This is a stance that insists on mutuality (neither perceiver nor the perceived causes the other). A distinction has emerged and with it a world with an observer and an observed. The distinction is itself an eigenform We identify the world in terms of how we shape it. We shape the world in response to how it changes us. We change the world and the world changes us. Objects arise as tokens of a behavior that leads to seemingly unchanging forms. Forms are seen to be unchanging through their invariance under our attempts to change, to shape them" [14].

Thus, in accordance with Kauffman's main thesis, the notion of a fixed object becomes a notion concerning the process that produces the apparent stability of the same object. This process can be simplified in a model to become a recursive process where a rule or rules are applied time and time again. The resulting object of such a process is the eigenform of the process, and the process itself is the eigen behavior. In other words, we have a model for thinking about objects as tokens for eigen behavior as advocated by H. von Foerster. In particular, Kauffman's model examines the result of a simple recursive process carried to its limit. For example, suppose that $X = (F(F(F\ldots)))$, that is, each step in the process enclose the results of the previous steps within a frame. These objects, these infinite nest of frames, may go beyond the specific properties of the world in which we operate. They attain their stability through the limiting process that goes outside the immediate world of individual actions. We need, in this sense, an imaginative act to complete such objects to become tokens for eigen behaviors. It is impossible to make an infinite nest of frames: as the great scholar remarks, we do not make it, we imagine it. And in imagining that infinite nest of frames, we arrive at the eigenform. "The leap of imagination to the infinite eigenform is a model of the human ability to create signs and symbols. In the case of the eigenform X with $X = F(X)$, X can be regarded as the name of the process itself or as the name of the limiting process" [15]. It is in the exercise of his imagination that Narcissus will be able to come into contact with the existence of his I until he himself is reflected and objectified in it at the level of the resulting tissue of the involved eigenforms. The image marks his advent to the truth and it is in this same image that Narcissus drowns. In this sense, the form of an object is the form of the distinction that it operates in the space of perception.

At the level of the ancient Myth, the Minotaur is initially pure magma (and as such he is blind, pure plot of intensities devoid of boundaries in himself). When the distinction is born (as eigenform) and with it the object, this means that the form of the distinction (which remains constant) has produced an object for the observer. The Minotaur succeeds in opening his eyes in the same time that an object comes to appear. Unity of the structures of perception, on the one hand, and of the objects seen, on the other hand. Reality brings into play a form of distinction that remaining constant produces, in turn, an object for the observer-operator: $O(A) = A$, where the object A constitutes an eigenform for the observer O. When the distinction emerges here is both perceiver and perceived, and here is a particular ability that draws existence along its own realizing itself according to a form in action. In this context, the Minotaur appears as a paradigmatic example of a non-trivial machine in the sense of von Foerster, i.e. of a self-organizing machine. In other words, at the level of the Minotaur, an eigenform constitutes the order for the process that generates it. Always according to the ancient Myth, Narcissus comes to existence by drowning in his image as a fixed point of self-replication. From a general point of view, the object is given by the process that determines the stability (invariance) of the object and this process is recursive. It is, however, every time, a certain type of self-organization that gives rise to the specific embodiment related to this type of process. This embodiment would not, in fact, be possible (as we have just seen) without the enchantment created by the flute played by Pan: the universe of intensities necessarily follows the *experimentum* and the irruption. Starting from Narcissus, moreover, there is no novelty that does not constitute a variation that arises from a Chance that represents, in turn, the other face of the Empire of Necessity. Here is the eternal spinning of the Wheel of Time in order to ensure through Omega (in accordance with Chaitin's deep intuitions) the perennial establishment of objectivity. Hence the remaining of the subject of perception identical to himself even in the presence of a creativity at play, a creativity, however, that must reveal itself as phantasmal (cf. Pier Francesco Mola, "Diana and Endymion", Rome, Musei Capitolini), i.e. as a creativity that tells of its reached invariance and for which life itself is reduced to such story-telling: i.e. to the inspection of itself as a fixed point. In this picture the recursive processes that ensure the relaxation of the imagination take, in some respects, the place of the ancient Kantian schemes.

Just as in Kant there are different types of schemes at work, also for what concerns the eigenforms we are faced with the blossoming of different types of eigenforms: boxes, frames, fractals, etc. In this respect, the real problem is the fact that Kauffman's model does not foresee the presence of specific semantic orderings at play, nor the intervention of the component related to meaning in action. From an effective point of view, at the level of cognition (and life), as we have already seen, the basic interweaving concerns the ever changing bond between reflection and simulation, i.e. between Narcissus and Marsyas. As a craftsman I must simulate possible worlds on my own skin so that the God can come to affect me. I must pose myself as a gridiron for the realization of specific concepts, only then will I be welcomed by the Muse: a creative thought will rise in me to the extent that I will come to be added. Just as thought is leavening in me, allowing me to transfigure, observation leads me to closing in a sphere, (a sphere, however, that, as shown by Pier Francesco Mola, can enclose creativity in itself, but only in a phantasmal way). I add myself as an observer and I come to be part of a Temple whose columns carry my name-symbol. Narcissus is constituted as a column and, therefore, as a replication process taking place at the level of Nature to the extent that he gives rise to the inscription of himself in the stone and to the reading-observation of his own form as a form of distinction and visibility. It is inasmuch as the hero comes to be perceived and to perceive that he comes into existence (Berkeley) thus coming to add himself to the very heart of the world of Necessity: he presents himself both as object and symbol. He stands as an object along the giving of an eigenform. The observer-operator during his acting can only leave the form of the object as invariant. In this sense L. Kauffman can only be aware of the necessity that: "... in the course of examining the concept of reflexivity we will find that the essence of the matter is an opening into creativity" [16]. This appears necessary if one wants to avoid a definitive closure in the universe of Narcissus and, therefore, in a form of creativity characterized in phantasmal terms. However, as we will see, without recourse to the simulation models it does not seem possible to outline an adequate (and not phantasmal) space for what is the process of self-organization that underlies cognitive procedures. The closure in the Narcissus world not only does not allow us to work for morphogenesis but does not even allow us to prepare what is necessary for the genesis of a renewed invariance, an invariance that cannot be solely considered as the fruit of a perennial cyclic activity (the cyclic resurrection of the hero as a flower).

At the theoretical level, the underlying goal of Louis Kauffman is to preserve, with regard to molecular Biology, the scheme relative to the central dogma as formulated by Monod (a scheme essentially linked to the first order Cybernetics) but with a view to its partial overcoming by means of the recourse to the methods proper to the second-order cybernetics. In his opinion, it is possible to achieve this overcoming making recourse to an extended application of reflexive domains able to ensure the due space for the full articulation of creativity and the continuous emergence of specific novelties. Only the success of such an attempt could lead us to the heart of a mathematically adequate theory (not only phenomenological) of that particular theoretical set-up relative to the enactive mind as proposed by Varela during his last years of life (although on largely phenomenological bases). Whilst, however, for Monod the creative evolution derives from the coupling between Chance and Necessity, for Kauffman who wants to open up to the methods of the second-order Cybernetics, the novelties must come to be born from within the system itself through an inner dialectic game such as that suggested by the functioning of that particular cellular automaton related to the Game of Life as outlined by J. Conway. However, as we will see, also an overcoming of this kind could only reveal in terms of a phantasmal "appearance" giving rise once again to a modest form of life such as that of Endymion: in fact, it will not be able to account for those bonds that weave together invariance and morphogenesis according to a real circularity, a circularity that finds its foundation first of all in the mathematics of non-standard and in that continuous passage by levels which appears inextricably linked to the arising of continuous changes at the level of the Semantics at work.

2.3 The Birth of a New Wittgensteinian Paradigm: A World of Capacities in a Computational Setting

After the death of Varela another volume was published in the wake (for certain aspects) of the book published in 1999. Also at the level of this volume a central role is played by J. Petitot and his school. The volume published in 2004 (cfr. Carsetti, A. (Ed.) (2004), *Seeing, Thinking and Knowing. Meaning and Self-Organization in Visual Cognition and Thought*, Dordrecht, [17]) is, partially, focused on an ambitious problematic: actually, according to the Editor the volume should not be limited only to the investigation of the genesis of the embodied mind, it was also aimed to reach an initial clarification of the action exerted by the embodied meaning. This volume sees, in particular, the presence of an illuminating work by K. J. O'Regan, E. Myin and A. Noe, "Towards an Analytic Phenomenology: The Concepts of "Bodiliness" and "Grabbiness". It is precisely this article that will open new horizons through its come to determine an important change of perspective with respect to the studies concerning the genesis of the human mind. Next to this article in the book there is also a chapter by A. Carsetti dedicated to a broad analysis of the reality of the embodied meaning.

The book allows us to see, among other things, how the legacy of Varela came to spread in the context of a variety of research areas. What appears very clear, however, once we analyze this spreading in details, is that the interest of many scholars working in the wake of Varela' s heritage is increasingly converging today on the Enactivism and the Theory of the embodied mind according to the lines of the already cited paper by Varela of 1990 (whose title is already particularly illuminating in itself). It is precisely in this framework that the lines of research carried out by O'Regan, Myin and Noe come to open up further horizons to the gaze. We are faced with a real breakthrough but marked in conservatively. As a matter of fact, the reality of embodiment is not called in question, the setting, however, in which this reality is to be realized appears extremely diversified and complex with respect to the mathematical and epistemological tools carried out by Varela. In order to understand the ultimate nature of this intellectual shift let us remind, first of all, that starting from the years 60 of the last century we have witnessed the birth of a structuralist theory as regards the nature of numbers and, in general, of the mathematical structures underlying human cognitive activity. This birth in many respects follows the progressive affirmation of the Computational Structuralism and appears strictly linked, as we shall see, to the consequences inherent in the Tennenbaum theorem.

It dates back to 1965 the publication of the fundamental work of Benacerraf (cf. "What numbers could not be" (1965), *Philosophical Review*, [18]). It is precisely on the wave of the theses advocated in this paper that the Computational Structuralism has recently developed. Hence the emergence of a very interesting debate that has come to focus on the problem relative to the genesis of mind in its link with the mathematical structures, a debate that has recently seen P. Quinon and K. Zdanowski come to revisit in 2006 the main thesis by O'Regan, Myin and Noe in order to outline a new vision of what was, in Kantian terms, the main tool available to our mind, namely that "Schematism" considered by Kant as the most mysterious (and perhaps inaccessible) aspect of mind's activity [19]. According to Kant the Schematism is the tool implemented at the level of human mind in order to guarantee the coordination of the concepts in presence of that self-organization process that identifies the genesis of the "Thinking I". The term "self-organization" is directly utilized by Kant in his major work (1781). As we have just mentioned, Kant's main intuition has been revisited by H. von Foerster at the level of the contemporary theory of self-organization, through the innovative recourse to the concept of non-trivial machines, of machines, that is, capable of self-organizing themselves. An interesting example of this kind of machines is today represented, albeit at a first level of complexity, by the associative memories as investigated by Kohonen in the framework of his theory of self-organizing maps [20]. The reference to the self-organization procedures acting at the cognitive level represents perhaps the most appropriate key to understanding the nature of the paradigm shift that animates the work carried out by Quinon and Zdanowski in the wake of the first insights by O'Regan, Myin and Noe.

Indeed, at the level of the paper by P. Quinon and K. Zdanowski (2006) entitled: "The Intended Model of Arithmetic. An Argument from Tennenbaum Theorem" we are faced with a shift that manages to field, beyond Husserl's lesson, the theses advocated by Wittgenstein in relation to his conception of meaning as use (as well as, in some respects, that theoretical perspective concerning the embodied meaning to which the last chapter of the aforesaid volume *Seeing, Thinking and Knowing* is dedicated). In order to understand the scope of this statement let us now to refer briefly to the lines of that scientific debate which, starting from 2004, finally came to focus on the reality of mind but with reference to the development of a Wittgensteinian perspective based on the theory of computation as well as on the mathematical instruments currently offered by the ongoing research in the field of hypercomputation. A reality this latter that appears far removed from the mathematical training by Varela. It is no longer a question of taking into account Turing's world together with the autonomous reality of a body according to the theses advocated by Varela in the aforementioned article he published in 1990, on the contrary the real issue now is to understand how from within the same reality of a computational self-organizing world that particular novelty represented by the evolution of a body together with its unique mind can come to be born. Once again, in view of being able to navigate in these mysterious waters the compass, in the first instance, can only be the one represented by the two theses by H. Atlan: (a) function self-organizes together with its meaning and A. Carsetti: (b) meaning self-organizes together with its creativity. That the genome cannot be identified as a simple program as claimed by Atlan, is quite clear in terms of the dynamics we can inspect at the deep biological level. As we have just said, biological information cannot be reduced to simple Shannonian information. Software identification, in turn, cannot be separated from the different phases of the self-organization process at work. At the level of the original library of programs it is necessary to identify both a depth dimension and a surface dimension;

moreover we are also witnessing the continuous construction of biological (non-trivial) machines and their incessant recycling and transformation. Hence the giving of a body of Nature as hardware forged in obedience to the "programs" at play on the basis of the selection exercised by the orderings offered by the Form, a Form that will finally come to reflect itself as pure software in action. Fundamental at this level is the action of editing as connected to the processes of splicing. We have, on the one hand, genes as "predicaments" (functional modules in action) working at the DNA level and, on the other hand, pure "insights" working at the level of the *Forma formans* considered as a source of individuated orderings. In this sense, the different intensities (capacities) can only be revealed in precise correspondence to specific sections of the functional library in action at the biological level. In such a theoretical framework, the original "irruption" can only appear linked to a language that becomes thought in life. Conception, on the other hand, appears to be linked to that thought through forms (operating in the framework of a vision in the truth) which becomes pure language, the language of a particular body finally drowning in the waters of the Temple. Hence the possibility of a self-reflection (mirroring) on behalf of the *Forma formans* in the body of Nature, a mirroring which finally presides over the birth of the Lord of the garlands and the triumph of the modules characterizing the artificial life. As we have just seen, it is exactly this kind of mirroring which identifies the role played by Narcissus. Meaning can reflect itself in the body of Nature to the extent that Narcissus comes to drown in his own image in waters.

The embodied life is memory plus imagination but along the realization of that particular detachment from the original *Sylva* as performed by the Minotaur. Here we can find the roots of Narcissus' development. From the intensities to the eigenforms trough the individuation of the correct reflexive domains. Only if the original memory self-organizes as pure hardware on the basis of the intervention of the due orderings will we be faced with new imagination and new data bases at work: only in this way true morphogenesis can, then, take place.

It is Narcissus "who" allows meaning to self-conceive and the source to realize the process relative to the self-renewing "inscription". He can really offer his severed head and his achieved ability as editor only to the extent to which he recognizes himself in the inscribed design and in the reflected image. Through the coder the source assumes a reproductive capacity commensurate with a precise invariance and with the individuation of intrinsic forms which inhabit life; it inscribes itself as instructional form (information) and as a hereditary principle in action, as a source of varied complexity but compared with a hereditary apparatus which self-organizes as such in view of possible regeneration. The source which generates on the basis of self-reflection (but at the surface level) opens out, then, towards a self-reproduction process which is targeted and part of a co-evolutionary path. Whoever arrests and captures the reflection, fixing and freezing it, also makes him/herself into a reflection; the offering of him/herself as severed head to the fluxes is in sight of a new invariance and the possible emergence of ever-new specific properties. Master/mistress of the shadow, s/he guides the process of regeneration by opening up to the new "conception". The inscription and the suture of the wounds operate at the level of the becoming body; and it is by a path of perceptual activity of this kind that a world articulated in properties is finally recognized. The result is a source which, having stored a pathway (the Road as this emerges along the petrifaction), is able to code and articulate as a set of properties and recipes, thus proposing itself as mirror to itself but within the contours of incoming life. Hence the possibility of a genomic information self-mediated in the architecture of proteins.

With respect to this mainframe, if we take into consideration, for instance, visual cognition we can easily realize that vision is the end result of a construction realized in the conditions of experience. It is "direct" and organic in nature because the product of neither simple mental associations nor reversible reasoning, but, primarily, the "harmonic" and targeted articulation of specific attractors at different embedded levels. The resulting texture is experienced at the conscious level by means of self-reflection; we actually sense that it cannot be reduced to anything else, but is primary and self-constituting. We see visual objects; they have no independent existence in themselves but cannot be broken down into elementary data. Grasping information at the visual level means managing to hear, as it were, inner speech. It means first of all capturing and "playing" each time, in an inner generative language, through progressive assimilation, selection and real metamorphosis (albeit partially and roughly) and according to "genealogical" modules, the emergent (and complex) articulation of the semantic apparatus which works at the deep level and moulds and subtends, in a mediate way, the presentation of the functional patterns at the level of the optical sieve.

What must be ensured, then, is that meaning can be extended like a thread within the file, a thread that carries the choices and the piles related to specific symmetry breakings. It is Narcissus who must donate cues in order to operate the fixing of meaning. At the end of the metamorphosis the hero will mirror himself in the motionless face of Ariadne. Now his head will be cut (cf. Caravaggio: Medusa's consciousness) and a vision will arise in accordance with the truth.

In this way, it will be possible to identify a "garland"; only on the strength of this construction can an "I" posit itself together with a sieve: a sieve in particular related to the world which is becoming visible. In this sense, the world which then comes to "dance" at the level of the eyes of my mind is impregnated with meaning. The "I" which perceives it realizes itself as the fixed point of the garland with respect to the "capturing" of the thread inside the file, inside, that is to say, that genealogically-modulated articulation of the file which manages to express its invariance and become "vision" (visual thinking which is also able to inspect itself), anchoring its generativity at a deep semantic dimension. The model can shape itself as such and succeed in opening the mind's eye in proportion to its ability to permit the categorial to anchor itself to (and be filled by) intuitions (which are not, however, static, but emerge as linked to a continuous process of deconstruction of the original holistic meaning). And it is exactly in relation to the adequate constitution of the channel that a sieve can effectively articulate itself and cogently realize its selective work at the informational level. This can only happen if the two selective forces, operating respectively within an ambient meaning and an ambient incompressibility, meet, and a *telos* shapes itself autonomously so as to offer itself as guide and support for the task of both capturing and "ring-threading". It is the (anchoring) rhythm-scanning of the labyrinth by the thread of meaning which allows for the opening of the eyes, and it is the truth, then, which determines and possesses them. Hence the construction of an "I" as a fixed point: the "I" of those eyes (an "I" which perceives and which exists in proportion to its ability to perceive (and "fix") according to the truth). What they see is generativity in action, its surfacing rhythm being dictated intuitively. What this also produces, however, is a file that is incarnated in a body that posits itself as "my" body, or more precisely, as the body of "my" mind: hence the progressive outlining of a meaning, "my" meaning which appears gradually pervaded by life. Concepts and schemes, intuitions and diagrams relative to orderings. Determinations of Time and diagrams of the memory. The intensities come to life, the Time comes to the truth. On the one hand schemes working on predicaments, on the other hand, orderings individuating the original insights. The categories through the schemes give rise to concepts, the Form through the orderings gives rise to intuitions. Determinations of Time and channeling of intensities. Conceptual acts and intuitive scans. The Form is filled with insights, the categorial, in turn, is populated by concepts.

Vision as emergence aims first of all to grasp (and "play") the paths and the modalities that determine the selective action, the modalities specifically relative to the revelation of the afore-mentioned semantic apparatus at the surface level but in accordance with different and successive phases of generality. These paths and modalities thus manage to "speak" through my own fibers. It is exactly through a similar self-organizing process, characterized by the presence of a double-selection mechanism, that our mind can partially manage to perceive (and assimilate) depth information in an objective way. Here we can find the ultimate roots of its genesis. The extent to which the network-model succeeds, albeit partially, in encapsulating the secret cipher of this articulation through a specific chain of functional modules determines the model's ability to see with the mind's eye as well as, in perspective, the successive irruption of new patterns of creativity. Only if the Minotaur manages to open his eyes, can Marsyas successively perform his *experimentum crucis*. To assimilate and see, the system must first "think" internally of its secret abstract "capacities", and then posit itself as a channel (through the precise indication of forms of potential coagulum) for the process of opening and anchoring of depth information. This process then realizes itself gradually into the system's fibers, *via* possible selection, in accordance with the coagulum possibilities and the meaningful connections offered successively by the system itself (as immersed in its meaning).

The revelation and channeling procedures thus emerge as an essential and integrant part of a larger and coupled process of self-organization. In connection with this process we can ascertain the successive edification of an I-subject conceived as a progressively wrought work of abstraction, unification, and emergence. The fixed points which manage to articulate themselves within this channel, at the level of the trajectories of neural dynamics, represent the real bases on which the "I" can graft and progressively constitute itself. The I-subject (the observer) can thus perceive to the extent in which the single visual perceptions are the end result of a coupled process which, through selection, finally leads the original Source to articulate and present itself as *true* invariance and as "harmony" within (and through) the architectures of reflection, imagination, computation and vision, at the level of the effective constitution of a body and "its" intelligence: the body of "my" mind. These perceptions are (partially) veridical, direct, and irreducible. They exist not only in themselves, but, on the contrary, also for the "I", but simultaneously constitute the primary departure-point for every successive form of reasoning perpetrated by the agent. As an observer I shall thus witness *Natura naturata* since I have connected functional forms at the semantic level according to a successful and coherent "score". In this sense at the level of the reflexivity proper to the system the eigenforms reveal themselves as an integrant part of that self-organization process which constitutes the real engine of visual cognition, a process that the *telos* itself can manage to "imagine" only along the progressive unfolding of its reflexive tools. Without the individuation of the "I" and the composition of the eigenforms no perceptual activity is really possible [21].

The multiple unfolding of the eigenforms will be tailored to the symmetry breakings that manage to be realized at the level of meaning. They come to constitute themselves as fixed points in the process of construction of the structures of the operator. In this sense, they present themselves as the real bases of my own perceptual operations and, therefore, "preside" at the identification of the objects in the world. The forms arise from the determinations of the embodiment taking place by means of the "infixions" offered by Ariadne. When the garland closes and embraces the Minotaur that embodies and reflects himself, we have the emergence of vision. Narcissus–Minotaur will finally be able to recognize himself as an invariant reality and a source of replication through his rendering to the "stone", i.e his realization as living being in the world and his becoming an integrant part of the ruler (through self-adjunction).

In accordance with these intuitions, we may tentatively consider, from the more general point of view of contemporary Self-organization theory, the network of meaningful (and "intelligent") causal "programs" living at the level of our body as a complex one which forms, articulates, and develops, functionally, within a "coupled universe" characterized by the existence of two interacting selective forces. This network gradually posits itself as the real instrument for the actual emergence of meaning and the simultaneous, if indirect, surfacing of an "acting I": as the basic instrument, in other words, for the perception of real and meaningful processes, of "objects" possessing meaning, aims, intentions, etc.: above all, of objects possessing an inner plan and linked to the progressive expression of a specific cognitive action.

As we have just said, the mechanism which "extracts" pure intuitions from the underlying formal co-ordination activity, if parallel to the development of the *telos* as editor with respect to the coder, is necessarily linked to the continuous emergence of new mathematical moves at the level of the neural system's cognitive elaboration, This consideration inviting the revisiting of a number of Kantian hypotheses. It would appear, for instance, to be necessary not only to reread Kant in an evolutionistic key (cf. e.g., K. Lorenz), but also with reference to other speculative themes like, for instance, the indissoluble link existing between life and cognition and between chance and necessity. Taking into consideration coder's action opens up a new and different relationship with the processes of mathematical invention, making it necessary, for example, to explore second-order territories, the very realm of non-standard mathematics as well as the dialectics between observer and observed reality [22].

Pace Kant, at the level of a biological cognitive system sensibility is not a simple interface between absolute chance and an invariant intellectual order. On the contrary, the reference procedures, if successful, are able to modulate canalization and create the basis for the appearance of ever-new frames of incompressibility through morphogenesis. This is not a question of discovering and directly exploring (according, for instance, to Putnam's conception) new "territories", but of offering ourselves as the matrix and arch through which they can spring autonomously in accordance with ever increasing levels of complexity. There is no casual autonomous process already in existence, and no possible selection and synthesis activity *via* a possible "remnant" through reference procedures considered as a form of simple regimentation. These procedures are in actual fact functional to the construction and irruption of new incompressibility: meaning, as *Forma formans*, offers the possibility of creating a holistic anchorage, and is exactly what allows the categorial apparatus to emerge and act according to a coherent "arborization". From the encounter of Ariadne with Narcissus we shall have the flowering of forms, the possibility to perceive by fixed points, and the birth of specific structures at the level of the operator.

The new invention, which is born then shapes and opens the (new) eyes of mind: I see as a mind because new meaning is able to articulate and take root through me. As J. Petitot correctly remarks, according to Kant the pure intuitions are: «"abstraites de l'action même par laquelle l'esprit coordonne, selon des lois permanents, ses sensations" (*Dissertation, 177)*. Or, cette coordination est elle-même innée et fonctionne comme un fondement de l'acquisition» [23]. In this sense, the space appears as a format, the very basis of spatial intuition is innate. According to Kant, it is a condition of a subject knowing anything that the things he knows should be unified in a single consciousness. Kant calls this consciousness the transcendental unity of apperception. Kant writes that this unity comes about "not simply through my accompanying each representation with consciousness, but only in so far as I conjoin one representation with another" (B 133, p. 153 in Kemp Smith) [24]. In this sense, all coherent consciousness, hence all knowledge of anything, presupposes not just an original unity, but original, conceptual *acts* of possible combination to produce such unity. This means that some concepts are *a priori*. They cannot possibly have been derived from experience, since without them there would have been no original unity of experience. Just as Kant identifies in this way the existence of *a priori* conceptual acts (by *coniunctio*) living at the level of the potential intellect, he also identifies, as we have just said, the existence of innate tessellations (by orderings) at the level of the Form. It is in dependence of the determinations operated by the schemes on the Form of Time (in connection with the operations of categories-capacities) that we will have the intuitions that populate our perception of the world.

The schemes allow a space-time tessellation according to intuitions. Here are the processes that lead to the articulation of specific eigenforms. From here, moreover, the profiling of that digital image in which Narcissus drowns reaching its objectivity. Starting from the capacities and the action of Grace we reach the eigenforms and the invariance. This, however, allows the new conception as well as the path pursued by the Work, the same come to burn in the air, in the close exploration of its most hidden secrets, by the Painter (De Nittis). Here is the current extroversion and the related eigenvalues. Hence the possible trigger and the new incarnation related to it (an incarnation that arises from irruption and not from conception). Capacities as intensities and as articulations of thought in life that are made, therefore, to eigenforms. On the opposite side, however, we are faced with meanings that are made to eigenvalues. Meaning as a trigger for the incarnation following the pyre, creativity as a condition for the abstraction following the petrifaction.

In this sense, at the biological level, as we have just seen, what is innate is the result of an evolutive process and is "programmed" by natural selection. Natural selection is the coder (once linked to the emergence of meaning): at this level the emergence process is indissolubly correlated to the continuous construction of new formats in accordance with the unfolding of ever new mathematics, a mathematics that necessarily moulds coder's activity. Hence the necessity of articulating and inventing a mathematics capable of engraving itself in an evolutive landscape in accordance with the opening up of meaning. As we have just said and as we shall see in the following chapters, the realms of non standard-models and non-standard analysis represent today a fruitful perspective in order to point out, in mathematical terms, some of the basic concepts concerning the articulation of an adequate intentional information theory. This individuation, on the other side, presents itself not only as an important theoretical achievement but also as one of the essential bases of our very evolution as intelligent organisms. Here we can find the basis for the development of the mathematical language related to a new science: Metabiology.

2.4 Beyond Pure Reflexivity: Meaning as a Trigger for the Incarnation

With respect to this theoretical mainframe, a reflexive domain should be considered not as an already-existing structure but as the actual *composition* of an existing structure with the "creative" exploration of the new possible higher-order informational paths stemming from it. Actually, a reflexive space is endowed with a non-commutative and non-associative algebraic structure. It is expandable and open to evolution over time as new processes are unfolded and new forms emerge. In a reflexive domain every entity has an eigenform, i.e. fixed points of transformations are present for all transformations of the reflexive domain. As we have just seen, according to von Foerster and L. Kauffman the objects of our experience are the fixed points of operators, these operators are the structure of our perception. In the process of observation, we interact with ourselves and with the world to produces stabilities that become objects of our perception. Our perceptual activity, however, is conditioned by the unfolding of the embodiment process and is linked to the cues offered by Narcissus to meaning in action. Moreover, these objects attain their stability through specific mathematical tools, through the unfolding, in particular, of specific limiting processes. Every recursion has a fixed point. Let us, however, point out that the notion of reflexive domain first appeared in the work of A. Church and H. Curry in the 1920s. Actually, the key to lambda calculus is the construction of a self-reflexive language: at this level we can solve the eigenform problem without the excursion to infinity. In the Church-Curry language the two basic rules: Naming and Reflexivity allow language to refer to itself and to produce itself from itself. In such a context, every object is a process and the structure of the domain as a whole comes from the relationships whose exploration constitutes the domain. As L. Kauffman remarks any given entity acquires its properties through its relationships with everything else. Hence the triumph of Reflexivity.

The reflexive models do not lead, however, to true creativity and real metamorphosis because they do not give an account for the emergence processes living at the level of meaning. In particular, these models do not take into consideration the dialectical pairing of incompressibility and meaning. They do not loosen the knot of the intricate relationships between invariance and morphogenesis and do not arise in relation to the actual realization of a specific embodiment. Hence the necessity of making reference to theoretical tools more complex and variegated as, for instance, non-standard mathematics and complexity theory. For instance, the von Koch curve is a eigenform, but it is also a fractal. However, it can also be designed utilizing the sophisticated mechanisms of non-standard analysis. In this last case, we have the possibility to enter a universe of replication, which also opens to the reasons of a limited emergence. At this level, the growth of the linguistic domain, the correlated introduction of ever-new individuals appears strictly linked to the opening up of meaning and to a continuous unfolding of specific emergence processes with respect to this very opening. However, in order to capture the reasons of a true (and full) emergence we are immediately faced with some specific imperatives: the need, for instance, for the introduction of precise evolutionary parameters, we have, in general, to bring back the inner articulation of the eigenforms not only to the structures of "simple" perception but also to those of intentionality. As we shall see, we shall be obliged to introduce a new theoretical landscape with reference to the traditional concept of complexity as it stands in accordance with the classical mathematical formalism in use. Actually, we need to get even deeper blacks of general semantics and conceptual complexity. We have, in particular, to explore the realm of non standard models in order to capture not only the reasons of invariance bur also the secret path of morphogenesis.

The universe of Reflexivity as delineated by L. Kauffman is the universe of invariance (by fixed points) which is given through imagination. For there to be imagination, however, it is necessary, as we have just said, to refer to the emergence of the Minotaur as well as to the encounter between capacities and orderings as mediated by Grace. The result will be given by the petrifaction (with the connected Assumption). We must necessarily refer, first of all, to the dialectic between creativity, on the one hand, and meaning, on the other. We are not only faced with observation (naming) and Reflexivity, but also with thought and extroversion. In particular, we need to be able both to feel the incisions made by the knife of the God on our skin and to invent new worlds starting from them. As we shall see in the following chapter, to do this we will need to climb on our shoulders. The passage from one observer to another, as outlined by T. Skolem, appears, in this sense, linked to the growth of creativity but through openness in the deep, through the inner articulation of naming as a function of the flare of meaning. It is only the trigger represented by the track related to the eigenvalues on the carpet as well as to the melody played by Pan that can open to Reflexivity and fixed points as these last will be configured at the level of observation and operative incarnation. Indeed, beyond the achieved invariance we have also to ensure the correct reasons in view of the rising of new morphogenesis.

In this sense, there is a circularity not only between observers and observers as advocated by L. Kauffman but also between observation and thought (and between observers and craftsmen as advocated by Prigogine). We are faced, in particular, with a language that becomes thought in life and that is articulated along the subsequent petrifaction by fixed points with reference to a reality intersected by observers and perceptual acts. This is precisely the reality of observation and not of thought. In this sense, the universe related to the flesh and the sensibility is expressed primarily on the observational level and not in direct reference to the revisable thought as correctly maintained by G. Kanizsa. On the horizon, however, of cognition in all its complexity there are not only eigenforms, but also eigenvalues. Here is a world that essentially refers to complex numbers and that is inserted in a context dominated by the laws of the Quantum Mechanics and, in perspective, by forms of computation that go beyond the same Quantum Mechanics. In this sense, reality cannot be considered, as L. Kauffman affirms, on the basis of an intuition by H. von Foerster, as a one who splits into two to return necessarily to itself with respect to a Time of repetition considered as the guardian of an absolute invariance. Indeed, next to pure thought there exists, in fact, observation and next to imagination there is, of necessity, the realm related to invention.

The offering of the stone on the basis of a targeted project gives rise to both the arrest of the activity of devouring by Saturn and the first channeling of the intensities. Hence the coming to appear of renewed conceptual syntheses and with them of minds that will come, therefore, to discover the language engraved on the stone arriving, in this way, to recognize and read themselves as different I in action: *Et in Arcadia Ego* (cf. N. Poussin, *Et in Arcadia Ego*, Devonshire Collection). It is the I who discovers itself as an inhabitant of *Arcadia* but in its own death (the shepherds of Guercino), an I that makes itself to a mind but placing itself as a memory to itself. Here is an I who recognizes itself in a sepulcher, in its sepulcher (cf. U. Foscolo), reaching its real autonomy in the light of meaning, of that meaning (the Goddess) that reveals itself to this same I as Grace in action. Its words will remain, there will be no loss of information or possible devouring, we shall be faced with a chain of memory (with drops of an universal memory) as well as with the giving of a cyclical reality: the "strange" evolution, that is to say, of Narcissus who returns eternally to himself as a flower. From this Reflexivity enclosed in itself and connected to the delineation of a Time understood as pure repetition, authentic novelty can not be born. Narcissus coming back cyclically to life is always identical to himself. Only the relationship of the function to its meaning and vice versa can, in reality, determine the emergence of true novelties but on the edge of Chaos and along the giving of a real evolution through incarnation and abstraction twined together.

Here are cellular automata which connect themselves and which are subjected to extroversion allowing the emergence through percolation of a specific "track" as engraved by the God who comes to speak at the level of a self-organizing process concerning meaning in action. In this way, a specific invention path will come to operate a targeted intervention on the ongoing recovery process on the basis of the achieved extroversion. Here is a meaning that will come to modulate the expression of original creativity by allowing the leavening of deep informational flows but in its coming to lead them back to an intentional project. Here is, in particular, Marsyas as craftsman acting as a control room for the appearance of deep emotions, the very recesses of life. Here is the necessary support for coming, then, to free imagination along the constitution of a mind. Hence the full realization of the detachment by the Minotaur and the emergence of an I and its consciousness: the I, that is to say, in which the unity of mind is identified: perception and apperception. It is only through the brain that a mind can come to look out and it is always through a mind that a brain can come to simulate itself. If there is no outcropping at the level of Nature, no extroversion can be given at the level of Work: from Idea to Nature. The offer of the track will allow the emergence of ever new deep information flows.

Only the shepherd who will pose himself as a sickle to himself (in his coming to question the old memories but in the presence of the loving assistance by the Goddess) will be able to assure his own return to creativity (depending on his coming to constitute himself as mind in action). In this way, he will come to "assist" and determine the process of gestation of new Meaning with consequent birth of the new Lord of the garlands: it is following this birth and the successive development of the way to abstraction that we can then resort to a real extroversion with subsequent irruption. From here the *Sylva* and the road pursued by the Minotaur–shepherd: but from here also the discovery of the sepulcher-memory, the reading by the shepherd of the language of the old memories (a language that he recognizes as language in the Truth). We are faced with an always renewed categorial that comes to be language through concepts (and writing), but under the assistance of the Goddess. Narcissus drowns in the language, Marsyas burns in the thought: language in the truth and thought in life. If the thought in life opens up to the new *Sylva* and the irruption, the language in the truth, in its turn, opens to the Temple and to the new conception. It is because Narcissus drowns in the language relative to his digital image that the Goddess can enclose (and reflect) herself at the surface of the tablet. The Goddess adorning herself all alone (as shown by Tintoretto and Titian in some of their paintings) gives rise to the new conception.

2.5 Natural Evolution Is the Necessary Landscape: Life as Evolving Software

As is well known, according to Darwin the driving principle of evolution is the 'survival of the fittest'. In some recent papers and books G. Chaitin has revisited the scientific status of evolution theory starting, in particular, from the following questions: is it possible to give a mathematical proof of natural evolution? is it possible to give an explanation concerning the ultimate reasons that lead the living systems to evolve? In order to give a first answer to these difficult questions the great scholar took his first moves from a revisitation of his earlier works on Algorithmic Information Theory (AIT) [25].

From a general point of view, traditional information theory states that messages from an information source that is not completely random can be compressed. Starting from this statement, Chaitin in the seventies outlined his general idea of randomness as well as the conceptual notion of complexity that is at the core of AIT: if lack of randomness in a message allows it to be coded into a shorter sequence, then the random messages must be those that cannot be coded into shorter messages. From a conceptual point of view, we can affirm that a string s is random iff H(s) is approximately equal to $|s| + H(|s|)$. An infinite string z is random iff there is a c such that $H(z_n) > n - c$ for all n, where z_n denotes the first n bits of z. Let us, now, consider a specific program p of length $|p|$. The probability that p will be generated by applying a random process of two random variables $|p|$ times is $2^{-|p|}$.

Let s be an object encodable as a binary string and let $S = \{s_i\}$ be a set of such objects s_i then the algorithmic probability P is defined by:

$$P(s) = \sum_{U(p)=s} 2^{-|p|} \tag{2.1}$$

$$P(S) = \sum_{s_i \varepsilon S} P(s_i) = \sum_{U(p)=S} 2^{-|p|} \tag{2.2}$$

$$\Omega = \sum_{U(p)\downarrow} 2^{-|p|} \tag{2.3}$$

Ω is the halting probability (with null free data). Ω is random and, in the limit of infinite computing time, can be obtained in the limit from below by a computable algorithm.

$P_{C'}(s)$ and $\Omega_{C'}$ can be defined for any (not necessarily universal) computer U′ by substituting U′ for U in (2.1) and (2.3). There are infinitely many programs contributing to the sum in (2.1), but the dominating term in the sum for $P(s)$ stems from the canonical program s*. It can be shown that there are few minimal programs contributing substantially to the sums (2.1), (2.2) and (2.3). Thus the probabilities to produce a specific object s as well as the halting probability can also be defined by taking into account only the canonical programs. Let s be an object encodable as a binary string and let $S = \{s_i\}$ be a set of such objects s_i, then the algorithmic probability P^* is defined by:

$$P^*(s) = 2^{-|s*|} = 2^{-H(s)} \tag{2.4}$$

Then we can state the incompleteness theorem of Chaitin.

5.1 Theorem *Any recursively axiomatizable formalized theory enables one to determine only finitely many digits of* Ω.

Let us, now, investigate more accurately the conceptual content of this theorem in the light of that complicated plot that ties together depth information and surface information from the point of view of the computability theory. Turing showed in 1936 that the halting problem is unsolvable: there is no effective procedure or algorithm for deciding whether or not a program ever halts. If we consider a formal axiomatic system as a r.e. set of assertions in a formal language, one can immediately deduce a version of Goedel's incompleteness theorem from Turing's theorem. Let us define the *halting set* $K_0 = \{<x, y>: \Phi_x(y) < \infty\}$. Then, we can rephrase the original Turing's theorem directly as: "The halting set K_0 is not recursive". Moreover, it is easy to show that, at the same time, K_0 is recursively enumerable.

In accordance with this conceptual track, we can formulate the (first) incompleteness theorem of Goedel in terms of recursive function theory. This result is due to Church and Kleene. In the proof diagonalization is needed to show that K_0 is not recursive.

5.2 Theorem *There is a recursively enumerable set* K_0 *such that for every axiomatizable theory T that is sound and extends Peano arithmetic, there is a number n such that the formula "*$n \neq K_0$*" is true but not provable in T.*

We also have the possibility to derive a new proof of the theorem resorting to Kolmogorov complexity. In this case, however, we have as a result different examples of undecidable statements.

The attempts of Church and Kleene as well as Post's version of Goedel's theorem not only show, in terms of recursive function theory, that formal axiomatic systems are incomplete but, in certain respects, they also give some hints in order to outline an information-theoretic version of Goedel's theorem, a version that will be given later by Chaitin. In this version we can find precise suggestions about the possibility of introducing effective measures of the information power of formal axiomatic systems. Let us consider, for instance, an *N*-bit formal axiomatic system *T*. Chaitin's revisitation of Goedel's incompleteness theorem affirms that there is a program of size *N* which does not halt, but one cannot prove this within the formal axiomatic system. On the other hand, *N* bits of axioms can permit one to deduce precisely which programs of size less than *N* halt and which ones do not.

Chaitin's version emphasizes very well one of the central aspect of the intellectual "symphony" realized by Goedel. Actually, at the heart of Goedel's incompleteness theorem is the derivation (outside *T*) of:

$$\text{Cont}\ (T) \rightarrow \neg \text{Ey Proof}_T(\text{y}, `\gamma') \tag{2.5}$$

(where T is a system based on a first-order language whose intended interpretation is over the universe of natural numbers).

Cont (T) is the sentence that affirms that there is no proof of $\underline{0} \neq \underline{0}$: $\neg$ Ey Proof$_T$ (y, '$\underline{0} \neq \underline{0}$') and the unprovability of the sentence γ is formally stated as $\neg$ Ey Proof$_T$ (y, 'γ').

The Formula (2.5) can be formalized in T if T is minimally adequate. In this sense (2.5) reveals itself as a theorem of T. Thus if T is consistent, Cont (T) is not provable in it. In other words: for any minimally adequate formal deductive system T, if T is consistent then its being consistent is expressible in T but unprovable in it.

γ is provably equivalent to Cont (T). Thus, Cont (T) shows itself as an undecidable. As such it can be added to T as an additional axiom. We obtain, in this way, a stronger acceptable system. In certain respects, one may consider Cont (T) as a totally informative message, a message consisting only of information that can be obtained no other way. All redundancy has been cancelled from it. It is determined by means of the explicit construction of an undecidable Π_1 formula: the fixed-point lemma associates with any formal system T, in a primitive recursive way, a formula which says of itself "I am unprovable in T".

Insofar as we manage to realize, by self-reflection on our own reasoning, that our logical and mathematical inferences can be formalized by a given formal system, we also realize that the self-reflection is itself part of our mathematical reasoning. In this sense, it is also at the basis of the effective construction of the undecidable. It is precisely by means of self-reflection that we can go onto infer the consistency of the system. However, the act of self-reflection must remain outside the final result of this conceptual construction. We can better realize now in which sense, according to Hintikka, depth information can be defined but, at the same time, it cannot be effectively computed. Actually, depth information can be thought of as surface information at infinite depth. In certain respects, we can simply affirm that it can be calculated by an infinite process during which one can never know how close one is to the final value.

It is in the moment in which the hero abandons himself to his Muse (but in the awareness of this and in his walking alone along the path of the Hades and the hallucinogens) that it will be possible for him to carry out those procedures of extroversion that will lead him (eternal Vermeer) to be welcomed by the God of creativity according to a renewed praxis of art. The new arising selection is inherent in his adjunction to creativity. As a result of the ongoing selection (with consequent break-in) we will therefore find ourselves before a new possible incarnation: a real and unheard novelty will come to support the actual emergence of the God.

It is worth underlying that Ω is Δ^0_2. We have also to remark that Ω is relative to the chosen universal machine U and thus to a particular coding of Turing machines that is used. As is well known, Δ^0_2 sets have a very natural computational characterization, based on the idea of computability in the limit; this is the notion of "trial and error predicate" as developed by Putnam [26]. In this sense, Ω can be represented by a trial and error predicate. Chaitin is indeed aware of the fact that Ω is computable in the limit: in his opinion, with computations in the limit, which is equivalent to having an oracle for the halting problem, Ω seems quite understandable: it becomes a computable sequence. From a more general point of view, as Chaitin and Calude directly affirm [27], Ω is "computably enumerable". We have, moreover, to recognize that, at the level of extreme undecidability, the incompleteness results arising from Ω are in no way the "strongest". Ω is just one among various undecidable sets.

The fact that Ω is unpredictable and incompressible follows from its compact encoding of the halting problem. Because the first n bits of Ω solve the halting problem for all programs of n bits or fewer, they constitute an "axiom" sufficient to prove the incompressibility of all incompressible integers of n bits or fewer. If we consider the axioms of a formal theory to be encoded as a single finite bit string and the rules of inference to be an algorithm for enumerating the theorems given the axioms, by an n-bit theory we can indicate the set of theorems deduced from an n-bit axiom. Remembering that the information content of knowing the first n bits of Ω is $\geq n - \mathrm{c}$ and that the information content of knowing any n bits of Ω is $\geq n - \mathrm{c}$, we are finally able to prove that if a theory has H (Axiom) $< n$, then it can yield at most $n + \mathrm{c}$ (scattered) bits of Ω. (A theory yields a bit of Ω when it enables us to determine its position and its 0/1 value).

On the basis of these conceptual tools and with reference to the afore mentioned questions, Chaitin outlines a first sketch of his model stating that life should be directly considered as evolving software. As Delgado correctly remarks, according to Chaitin: "a living organism it is a classical program, i.e., a piece of software that can be fed in a universal Turing Machine and produce a certain output, or just halt or even not halt. If the program **O** halts, then the output is a string of classical bits x. In the theory of classical computation, a program **O** can also be characterized by a certain bit-string whose size is denoted as |**O**|. Thus, $\mathbf{O} \in \mathbf{\Pi}$.

The rationale behind this choice is an abstract process that reduces an organism to pure information encoded in its DNA. The rest of the organism such as its body, functionalities etc are disregarded as far as being essential to evolution is concerned" [28].

If a living organism can be seen as a classical program, evolution is to be considered as a random walk through a software space. The conceptual complexity H(M) of the mutation M is the size in bits of the program M. This is the key idea utilized by Chaitin in order to model Darwinian evolution mathematically. We are faced with an abstract process that reduces an organism to pure information encoded in its DNA, to instructional information, that is to say, searching for the best way to its self-organization. As we have just said, the rest of the organism such as its body etc. is disregarded. Let us emphasize, however, that the "inquiry" performed by the system is not an end in itself: it may constitute the instrument able to trigger new and more sophisticated levels of embodiment by changing Semantics through the "intelligent" recourse to the tools of non-standard mathematics. Here we can recognize one of the most fundamental aspects of that particular circularity that characterize the evolution of life. It is in the new emerging levels of the embodiment (as they will come to intersect the ongoing surfacing process) that the novelties relating to a real recovery of creativity will gradually come to graft themselves.

Chaitin's deep insight into the problem of biological evolution is the choice of the fitness function with respect to the formalism provided by AIT. "The idea is to see life as evolving software, such that a living organism is tested after a mutation has occurred. The idea is to use a testing function that is an endless resource. This way, evolution will never be exhausted, will ever go on. In AIT, there are several functions with this remarkable property that make them specially well-suited for this task: quantities that are definable but not computable. One example is the Busy Beaver function **U**. Another example is Chaitin's Ω number" [29]. In Chaitin's model (2010), the Busy Beaver function exactly embodies a type of pressure. It represents the simplest possible challenge to force our organisms to evolve and to manifest their creativity. Using the Busy Beaver function, the fitness of a new program can be compared to the fitness of the previous program. If the new program appears as fitter it takes the place of the old program. The Busy Beaver function of N, BB(N), that is used in AIT is defined to be the largest positive integer that is produced by a program that is less than or equal to N bits in size. BB(N) grows faster than any computable function of N and is closely related to Turing's halting problem, because if BB(N) were computable, the halting problem would be solvable. As Chaitin remarks, doing well on the Busy Beaver problem can utilize an unlimited amount of mathematical creativity. For example, we can start with addition, then invent multiplication, then exponentiation, then hyperexponentials, and use this to concisely name large integers. According to Chaitin: "There are many possible choices for such an evolving software model: You can vary the computer programming language and therefore the software space, you can change the mutation model, and eventually you could also change the fitness measure. For a particular choice of language and probability distribution of mutations, and keeping the current fitness function, it is possible to show that in time of the order of 2N the fitness will grow as BB(N), which grows faster than any computable function of N and shows that genuine creativity is taking place, for mechanically changing the organism can only yield fitness that grows as a computable function" [30]. Subsequently, Chaitin (2013) was also able to show that in time of the order of $N^{2+\varepsilon}$ the fitness will grows as BB(N). In order to keep evolution from stopping, we stipulate that there is a non-zero probability to go from any organism to any other organism: $\log_2$ of the probability of mutating from A to B defines an important concept, the mutation distance, which is measured in bits. As we have just said, the conceptual complexity H(M) of the mutation M is the key idea utilized by Chaitin in order to model Darwinian evolution mathematically. We need a sufficiently rich mathematical space to model the space of all possible designs for biological organisms. In Chaitin's opinion (2013), the only space that is sufficiently rich to do that is a software space, the space of all possible algorithms in a fixed programing language. Here, we can find a correct mathematical image for the aforementioned space concerning the regulatory logic underlying cellular life. Chaitin's choice concerns an abstract process that reduces an organism to pure information encoded in its DNA. With respect to this frame of reference, the paper by Zenil et al. introduces some interesting improvements [31]. Zenil et al. correctly underline that, from a general point of view, computation may be considered an important driver of evolution: in particular, the results of the research indicate an accelerated rate when mutations are not statistically uniform but algorithmically uniform. The paper shows that algorithmic probable mutations reproduce (appear able to capture) aspects of evolution, such as convergence rate, genetic memory and modularity.

In a nutshell, Chaitin's ultimate goal is to prove mathematically that evolution through random mutations and natural selection is capable of producing the complexity and diversity of life-forms that populate our planet. This is the starting point concerning his project for Metabiology. Metabiology, in his opinion, is not about randomly mutating DNA but about randomly mutating software programs. As Virginia M. F. G. Chaitin remarks, it's a metatheory not a biological theory. In this sense, exactly as it happens for what concerns the definition given by L. Kauffman of Reflexivity, also at the level of the first draft by Chaitin of Metabiology, an essential role is played by creativity, by the emergence of ever new novelties to evolutionary level. Actually, the fitness as introduced by the great scholar in his metabiological model essentially refers to the creation of novelty in the biological realm. As Virginia Chaitin notes, according to Chaitin: "… we're not talking about creativity for finding out solutions for problems. We're talking about creativity about expanding the mathematical realm, expanding mathematical knowledge. So it's a creativity that brings about novelty within the mathematical realm. And metabiology posits a connection between creativity in mathematics and creativity in biology" [32]. In this context, it cannot be a surprise that for Chaitin mathematics appears to possess a quasi-empirical character to the extent, in particular, that it sometimes progresses by doing experimental math instead of searching for proofs.

If through extroversion procedures we succeed in identifying, with reference to a suitable software space, the plot relative to natural evolution in the framework of a random walk we have the opportunity to experience a real selection in view of a possible optimization. The problem, however, is that this should not be considered as an end in itself: in fact the aforementioned identification is first of all in view of realizing a new incarnation as well as an adjunction of the subject as renewed creativity. The creation of the novelty (the adjunction, that is to say, of Marsyas to himself on the part of the God with the consequent transformation in the bush of the artificial into the natural) does not take place for a *fiat* or for the effect of a Game (as suggested by Conway). The extroversion is really in function of the activation of new selection procedures capable of ensuring a metamorphosis (from within) of the system, a metamorphosis that will necessarily involve the mathematician who builds the model. Here is the one who along the abstraction process does not limit himself to identify the correct software space for an evolutionary trend already defined, but which, instead, comes to design a software space capable of ensuring the emergence of a new possible embodiment. It will therefore be necessary both to think in terms of new semantics and to preside over specific *experimenta* with reference to oracles and so on. The novelty that will be born will now inhabit a space that can only be that of the new incarnation, the space of a renewed Nature that will be revealed crossed by a new and unprecedented language. From here it will only come to follow new extroversion.

According to Chaitin, the process we are now considering is not just darwinian evolution, it first appears as creativity in action on a metabiological level. In this sense evolution appears as a hill-climbing random walk in software space. As Chaitin states "Metabiology is a field parallel to biology dealing with the random evolution of artificial software (programs) rather than natural software (DNA)" [33]. We are making biology mathematical at a meta-level. From a general point of view, according to Chaitin, the goal of Metabiology is to find the simplest pythagorean life-form that has hereditary information and evolves according to a fitness criterion. According to the great scholar, we can maintain that Nature is in some way programming our metabiological organism, without a programmer though, because we don't need a programmer for metabiological evolution to take place [34]. Nature is programming without a programmer! In this respect, let us note that also according to Kant Nature is the result of a self-organization process. Marsyas applies his simulation at the level of the possible worlds not in order to create a parallel evolution rivaling the God but to offer himself as a tool for the channeling of new emotion. Here is a God who will be able to make the due recovery through the work performed by the Silenus. The moment in which Marsyas comes to be added is also the moment in which he succeeds in threading the garlands according to the truth thus allowing the Muse to come and sing in him and for him. Here is the rising of a song related to a History that is the story of humans and Gods, the song, in fact, of Clio as Muse of History.

In this sense, we must as humans be aware that the computational praxis related to the biological organisms (the autonomous agents as outlined by S. Kauffman, A. Carsetti etc.) constitutes the "abstract" support for their embodiment even if this same praxis makes these organisms (once disembodied) amenable to being studied as computer programs following random walks in software space. Actually, if we remove the embodiment process the situation looks like crystallized at the level of the (pure) realized abstraction: we are faced with an instructional information (the DNA as the volume of the instructions) searching for the best way to its self-organization. To paraphrase a famous sentence by Galilei, we can say that Nature speaks by computations. Doing this Nature is embodied. However, in view of the embodiment, natural computations cannot be articulated only on a purely syntactic level nor can they be flattened on disembodied crystallizations: as H. Atlan, A. Carsetti, S. Kauffman etc. state, at the biological level we are always confronted with semantic information and semantic phenomena concerning the ongoing self-organization. In this sense, Shannon's information cannot be directly applied to Biology: the same extension proposed by Zenil et al. to a context-sensitive formulation of the original theoretical approach by Chaitin constitutes an important step in the direction of an adequate understanding of the natural behavior of the autonomous agents (the biological organisms) in the line of what has now been affirmed, a step, in any case that cannot appear as sufficient. Without adequate software, you cannot achieve true autonomy at the level of the embodiment, but without the realization of such autonomy software cannot come to identify and promote those particular constraints that alone can make the original "blind" information to life thus allowing the channelling of the original intensities in accordance with an intelligent opening up from within of the different levels of deep information. An opening that will be articulated through the utilization of new languages and new emerging functions. It is only with reference to this coupled context that a growth of creativity can arise on a natural level. The simple exploration of new possible software spaces does not necessarily lead to the emergence of true novelties. Moreover, the extension of Chaitin's approach cannot be limited to the context-sensitive component alone. It is necessary, for example, to take also into account the openings from within as they emerge at the level of the dialectic between codifying DNA and non codifying DNA, a dialectic that unfolds on the basis of the regulation put in place as well as of the work relative to the editing procedures. Here is the splicing system in action and here are the new functions that also emerge as a result of the enrichment of the software. In this sense, the coupled system cannot be split in itself if not at the modeling level: indeed, we shall be obliged to continuously compare the results obtained with the reality of natural evolution. If the function is to open itself to the level of its same depth in accordance with the evolution underway and if the autonomous agent proves capable of climbing on his own shoulders (through the tools of non-standard mathematics and through the recourse-invention to a constantly renewed Semantics), here is, of necessity, the coming to gush from the inside (from the bottom) of new constraints. Here is information *via* constraints as mentioned by S. Kauffman (2009) and A. Carsetti (1987, 2000). Hence the necessary trial with a canvas that is always unfinished and which, however, through the intervention of the appropriate software, can gradually reach ever higher degrees of circumscribed autonomy.

2.6 Limitation Procedures and Non-standard Models

From a general point of view, the limitation theorems are theorems that are based on a precise distinction between theory and metatheory, between language and metalanguage. A formal system can be considered as an "objectified" language and we well know that by means of precise arithmetical procedures the syntactical properties of a given formalized theory T can be expressed in terms of arithmetic predicates and functions.

We have just remarked, for instance, that Goedel's incompleteness theorem concerns a sentence of **Z** (where **Z** is a formal system obtained by combining Peano's axioms for the natural numbers with the logic of type theory as developed in *Principia Mathematica*) which says of the sentence itself that it is not provable in **Z**. However, the existence of such a sentence can be identified only because we are able to arithmetize metamathematics: i.e., to replace assertions about a formal system by equivalent number-theoretic statements and to express these statements within the formal system.

In this sense limitation theorems show that particular reality (or "essence") represented by "arithmetical truth" is not exhausted in a purely syntactical concept of provability. From a more general point of view, we can directly affirm that in **Z** we cannot define the notion of truth for the system itself. In other words, by constructing a system and then treating the system as an object of our study, we create some new problems, which can be formulated but cannot be answered in the given system. Actually, every sufficiently rich formal system is always submitted to the diagonal argument, an argument that is always present in the limitation theorems and, in particular, in the Löwenheim–Skolem theorem. Let us show, as a simple consequence of this last theorem, how one can prove that no formalized set theory can give us all sets of positive integers. Let S be a standard system of set theory. Since we can enumerate the theorems of S, we can also enumerate those theorems of S each of which asserts the existence of a set of positive integers. Let us consider now the set J of positive integers such that for each m, m belongs to J iff m does not belong to the mth set in the enumeration. By Cantor's argument, J cannot occur in the enumeration of all those sets of positive integers which can be proved to exist in S. Hence, either there is no statement in S which affirms the existence of J, or, if there is such a statement, it is not a theorem of S. In either case, there exists a set of positive integers which cannot be proved to exist in S.

In other words, the axioms of our formal system cannot give us a representation of all sets of positive integers. It is precisely in this sense that the systems containing these axioms must necessarily admit non-standard models.

Thus, the limitation procedures permit us to identify the boundaries of our intellectual constructions, to characterize, for instance, as we have just remarked, the class of natural numbers. They permit us to "see", once given a specific representation system **W**, that if **W** is normal then every predicate H (the predicates, in this particular case, can be thought of as names of properties of numbers) has a fixed point. They also permit us, for example, to identify an unlimited series of new arithmetic axioms, in the form of Goedel sentences, that one can add to the ancient axioms. Then, we can use this new system of axioms in order to solve problems that were previously undecidable.

We are faced with a particular form of mental "exploration" that, if successful, embodies in an effective construction constraining the paths of our intellectual activity. This exploration concerns the identification of new worlds, of new patterns of relations, the very characterization of new universes of individuals. We shall have, as a consequence, the progressive unfolding of an articulated process of cancellation of previously established relations and the birth of new development "languages" that are grafted on the original relational growth. As we have said before, this type of mental exploration articulates at the second-order level: it can be reduced however (if successful) at the level of many-sorted first-order logic, by means of well known logical procedures.

In a nutshell, the nucleus of this kind of reduction consists in explicitly showing in many-sorted structures what is implicitly given in second-order or in type theory. According to Post's famous thesis, any law we become completely conscious of can be mechanically constructed. So, we add to the many-sorted language membership relation symbols and to the many-sorted structures membership relations as relation constants. Throughout this reduction process, we simply consider that a second-order structure (or a type theory structure) is basically a peculiar many-sorted structure, since it has several domains. In short, we prove first of all that Henkin semantics and many-sorted first-order semantics are pretty much the same. Then, *via* Henkin semantics, we establish a form of reduction of second-order semantics to first-order semantics. Second-order logic with the Henkin semantics is, in general terms, a many-sorted logic.

However, we immediately have to emphasize that this kind of reduction does not imply that the secret "reasons" that guide, from within the mental activity, the progressive unfolding of the processes of exploration and invention can be reduced to a first-order mechanism or to a set of pre-established rules.

It is true that insofar as the aforementioned exploration process manages to embody in an effective construction that acts as a bunch of constraints and classification procedures, then we have the possibility to translate this kind of structure in a many-sorted language. But, the actual unfolding of abstract procedures that constitutes the primitive nucleus of the exploration process necessarily articulates (at least) at the second or higher-order level. As a matter of fact, the first result of this very unfolding is the birth of specific (and previously unknown) differentiation processes, as well as the successive appearance of new universes of individuals.

Let us quote Goedel, "P. Bernays has pointed out on several occasions that, in view of the fact that the consistency of a formal system cannot be proved by any deduction procedures available in the system itself, it is necessary to go beyond the framework of finitary mathematics in Hilbert's sense in order to prove the consistency of classical mathematics or even of classical number theory. Since finitary mathematics is defined as the mathematics of *concrete intuition*, this seems to imply that *abstract concepts* are needed for the proof of consistency of number theory ... By abstract concepts, in this context, are meant concepts which are essentially of the second or higher level, i.e., which do not have as their content properties or relations of *concrete objects* (such as combinations of symbols), but rather of *thought structures* or *thought contents* (e.g., proofs, meaningful propositions, and so on), where in the proofs of propositions about these mental objects insights are needed which are not derived from a reflection upon the combinatorial (space-time) properties of the symbols representing them, but rather from a reflection upon the meanings involved" [35].

In this sense, there must be proofs that are not fully formalizable at a given stage in our mental experience, but that are "evident" to us at that stage on the basis of particular arrangements of limitation procedures, of the successive identification of fixed points, of the utilization of abstract concepts, of the exploration of new universes of individuals, and so on.

In other words, there are, for instance, proofs of Con (PA) (primitive arithmetic) that require abstract concepts as well as the necessary construction of new elements, concepts, for instance, that are not immediately available to concrete intuition (Hilbert's concrete intuition as restricted to finite sign-configurations). We need, in general, not only rules, but also rules capable of changing the previously established rules. In Gödel's consistency proof, for example, we can directly see that the theory of primitive recursive functionals requires the abstract concept of a "computable function of type t".

Thinking in mathematical terms cannot be completely constrained within the boundaries of the syntax of a specific language. In fact, we would also need to know that the rules of this particular syntactical system are consistent. But in order to realize this, we will need, by the second incompleteness theorem, as we have seen before, to use mathematics that is not captured by the rules in question.

According to Gödel, utilizing mathematical reason we are capable of outlining and, at the same time, discovering specific abstract relations that live at the second-order level and that we utilize and explore at that stage in order to define first-order relations. We are faced with a particular "presentation" of the Fregean *Sinn*, a presentation that constrains the paths of our reasoning in a significant way. So, abstract and non-finitary intellectual constructions are used to formulate the syntactical rules. Once again, this is for many aspects a simple consequence of incompleteness results: mental constructions cannot be exhausted in formal concepts and purely syntactical methods. We have, in general, to utilize more and more abstract concepts in order to solve lower level problems.

The utilization at the semantic level of abstract concepts, the possibility of referring to the sense of symbols and not only to their combinatorial properties, the possibility of picking up the deep information living in mathematical structures open up new horizons with respect to our understanding of the ultimate nature of mental processes.

We are actually dealing with a kind of categorial perception (or rational perception) that does not concern simple data (relative to the inspectable evidence), but complex conceptual constructions. And we know that in Husserlian terms, meaning "shapes" the forms creatively. However, we must immediately remark that categorial perception appears to embody in a realm that is far beyond the limits of Goedel's primitive suggestions, in particular of his primitive Platonist approach. Actually, at the level of the articulation of mental constructions, we are faced with the existence of precise forms of co-evolution. On the one hand, we can recognize, at the level of the aforementioned process of inventive exploration, not only the presence of forms of self-reflection but also the progressive unfolding of specific fusion and integration functions, on the other hand, we find that the *Sinn* that embodies in specific and articulated rational intuitions guides and shapes, in a selective way, the paths of the exploration. It appears to determine, by means of the definition of precise constraints, the choice of some privileged patterns of functional dependencies, with respect to the entire relational growth. As a result, we can inspect a precise spreading of the development dimensions, a selective cancellation of relations and the rising of specific differentiation processes. We are faced with a new theoretical landscape characterized by the unfolding of a precise co-evolution process, by the presence, in particular, of specific mental processes submitted to the action of well specified selective pressures, to a continuous "irruption" of depth information determining the successive appearance, at surface level, of specific *Gestalten* [36]. This irruption, however, could not take place if we were not able to explore the non-standard realm in the right way, if we were not capable of outlining adequate non-standard models and continuously comparing our actual native competence with the simulation recipes. Selection is creative because it determines ever-new linguistic functions, ever-new processing units which support the effective articulation of new coherence patterns. And, it is precisely by means of these new patterns that we shall be able to "narrate" our inner transformation, to become aware of our mental development and, at the same time, to ascertain the objective character of the transformation undergone.

We can perceive the objective existence of abstract concepts only insofar as we transform ourselves into a sort of arch or gridiron for the articulation, at the second-order or higher-order level and in accordance with specific selective procedures, of a series of conceptual plots and fusions, a series that determines a radical transformation of our intellectual capacities. It is exactly by means of the actual reflection on the new-generated abstract constructions that I shall finally be able to inspect the realization of my autonomy, the progressive embodiment of my mental activities in a "new" unitary system.

Meaning can selectively express itself only through: (a) the determination of specific patterns of coherence; (b) the nested realization, at the conceptual level, of specific "fusion" processes; (c) a co-operative articulation of the primary informational fluxes. It shapes the forms in accordance with precise stability factors, symmetry choices, coherent contractions and ramified completions. We can inspect this kind of embodiment, at the level of "rational perception", insofar as we are capable of constructing and identifying the attractors of this particular dynamic process. It is with reference to the successive identification of these varying attractors that we can reach an effective self-representation of our mental activities. A representation that exactly concerns the "narration" relative to the *Cogito* and its rules.

2.7 The Dialectic Between Meaning and Creativity and Post's Machines in Action

Bacchus must fly high in the sky in order to welcome Arianna (assumed in Heaven) and make her his bride, the queen of the Kingdom of Culture. Hence the opening of the seething mantle of intensities and the new categorial that shines. Here is the fire hidden in the night of intensities, a fire devoid of any possible vision from which the emerging Minotaur will, therefore, come to slip away along his giving rise to the birth of new imagination. Here is a meaning that shows itself able to self-organize together with its creativity. It will be Bacchus who will come to the cross in the footsteps of Marsyas. He will come to experience the abandonment and will act as real support for the overcoming on behalf of the God of the self-abandonment of himself. It is starting from the recovery operated that, therefore, the irruption will arise and with it the new volcanic mantle of the intensities from which the new Minonaur will soon emerge. Marsyas has been added as a creator, thus bringing to completion, along his own metamorphosis, the self-organization of meaning.

As Marsyas teaches, it is necessary to fly into the simulation sky so that new eigenvalues can come to be individuated and defined in the very death of the Silenus, those eigenvalues, in particular, which will allow the God to make the appropriate selection along his come to sink the knife in the bark of the Silenus. Only the God who engraves properly, with reference, that is to say, to the correct eigenvalues, can, then, emerge in accordance with the stumbling in his own body by the Minotaur along the embodiment process.

Hence the giving of the body of Nature but written in mathematical characters. It is only the simulation work put in place as it is based on a continuous activity of invention that can lead to the constitution of Nature according to the truth, in agreement, that is to say, to the denial by Narcissus in meaning. Here is the relationship between function (creativity) and meaning and between categorial and Form. Information becomes semantic (and intentional) only through the metamorphosis, creativity, in its turn, is only achieved in the recovery: everything is tied to a continuous self-organization process.

It is necessary to fly high in the space of software so that new incarnation may come to be, but it is also necessary that the hero who simulates (acting as a craftsman) comes to individuate the path to the new incarnation through his own resurrection. This implies that we cannot close ourselves in Ω: at the same time, however, we cannot avoid referring to it. It is necessary to operate the disembodiment in the correct (and controlled) way so that a new inventive route to the incarnation may arise. Here is the need for the continuous exploration of Ω as postulated by Chaitin. It is the descent into the underworld that alone can lead to the new embodiment. This means that we will have to invent-build ever new mathematics, providing to continuously change Semantics but on our own skin: here are the successive metamorphoses of Marsyas in the different observers and here is the continuous rising of the Silenus on his own shoulders. Here is the hero of the simulation that resurrects: it is his gaze that really petrify the ancient remains. We will not only have new materials and prosthesis: it is also a new way of seeing and thinking what will come to be born. We'll find out as new to ourselves and no longer dressed, at the limit, of flesh alone and, yet, even more unpredictable. It is, in fact, our unpredictability that we must, above all, come to build at the level of natural evolution.

By identifying omega we have the possibility of going further, but this going beyond can only come to coincide with the progressive constitution of a new body and a new thinking in the overcoming of Ω itself. Here is the *bricolage* at work as well as the need for new simulation: here is the statue by Picasso ("Reading Woman", Musée Picasso) as an example of a mechanically assembled set up (an example of pure *bricolage*) yet able to express a reading of itself according to its coming to reflect itself in the stone book (articulated by eigenforms at the natural level) that opens up before its eyes. The statue is enclosed in its space (it constitutes the final result of the action of the *docta manus*) but it enters into communication with the observer opening up to the discovery of horizons not certainly present in its constitutive materials. Here is a meaning that becomes a function, which fills the materials utilized by Picasso with itself, making them evocative, thus introducing the craftsman-sculptor to a reading and an in-depth questioning of himself. We need to be able to work at the level of genetic engineering, i.e. to govern the deep information flows which intersect our body in the same time that we lower ourselves within them in order to facilitate their emergence. In other words, we have to prepare the tools to promote what will be revealed as completely new and which will also come to forge our own going beyond ourselves. The novelties will have to come into being in me, at the level of my imagination and my invention: we will therefore be obliged to make recourse to the dialectic between Marsyas and Narcissus in order to account for realities that will come to invade us disarticulating our original being.

The truth by Bernini is a body that ascends but as enlightened by God and by his resurrection. Here is the city that raises (cf. U. Boccioni) but based on eigenvalues that tell the story relative to the death of the Silenus and that reflect in themselves new ways to the computational thought never before taken into consideration. In that city lives a biological memory, in it we can find the tools for the realization of our future but in the metamorphosis of ourselves. Here is an imagination that comes to spring from an irruption, from the coming to explode by specific systems that open up to new intensities and new capacities. It is in the frame of this dialectic that a new role for the "thinking I" can come to be defined and along with it even those that are the prerogatives of consciousness will come to find their best definition with respect to the unfolding of a constantly renewed intentionality. Here we can find the same origin of the shelling away of the Minotaur in his coming to operate the detachment. Here is the coming to open again of new categorial as outlined by Kant but related to that secret key, unknown to Kant, represented by natural evolution, a key that lies within the recesses of the dialectic relationship that ties together abstraction and incarnation.

Here is an invention that brings reality back to "machines in action" (Post) in order that a new burst may take place: this is the ultimate meaning of the observer's going up on his/her own shoulders as predicated by Skolem. Hence, extroversion, Turing, and artifice. The journey to the underworld must, in any case, be supported by the inheritance of Narcissus: Reflexivity must guarantee adequate support for the journey into the realm of hallucinogens. The reality of the artificial, once it has entered the bush, will therefore give rise to the new incarnation, an incarnation that follows the irruption by the God as pervaded by a thousand emotions. But the artificial must reveal-prove able to capture in itself the whole flowering related to the productive imagination which had come to express itself on a mental level, an imagination that is linked to the emergence of the original emotions and, therefore, to the constitution of Nature. In this context, invention appears linked to the nesting of the rule (of the Goddess) and, therefore, to the establishment of the works. Once we have imagination that flows into Nature (and the body-incarnation) *via* the eigevalues and the primer operated by Pan and once we are faced with invention that leads to the Work and the soul-abstraction *via* the eigenforms and the seam of the garments by the Muse-model The emotion must be corrected according to the rule if we want to ensure a subsequent intentional extension by the emotion itself and, therefore, through this, the achievement of a true invariance, albeit in the change. The rule, in turn, must be nourished by emotion if we wish to ensure a full development of the rule itself and, therefore, a real morphogenesis, albeit in continuity. Here is Life and Truth coupled, and with them the Way, albeit in accordance with its two directions. Hence the limit but also the value of the theses advocated by Chaitin:

the great scholar is unable to grasp all the values of the role played by meaning at the level of evolution, but carries out an analysis of the artificial methods connected to the possible emergence of new natural processes. Undertaking the journey to the Columns of Hercules, Chaitin really sets the conditions for a paradigm shift, thus opening up (albeit *in nuce*) to a new Semantics and the new cries of future detachment. The journey into invention begins here, and it is precisely here that the role played by the verses takes shape at the beginning. No longer pure determinations of the Form but tools capable of inventing the very way of articulating things to say and think from within: that is, tools that open up to the new arising function. The extroversion and semantic clarification represent the first step in view of the irruption taking place. It is necessary, however, to feed the omega-related pyre in order to be able to ensure the correct modalities for the change of semantics: in other words, to be able to truly face that onerous passage constituted by the passage on one's shoulders by the hero as suggested by Skolem. The artificial must be reflected in itself and must reflect the imagery in place, with a view to preparing for a correct break-in. It is in this sense that, by linking Goedel, Turing, and Darwin, Chaitin offers a versatile and important contribution to that complex analysis that right now is progressively preparing the first foundations of a new science: Metabiology. Hence the first emergence of artificial but not trivial machines as imagined by von Foerster: biological machines able to self-organize and to stay in symbiosis with man in view of his becoming a new creator through his coming to be added to himself by the God (in accordance with Bergson's metaphor, later taken up by Monod). These are unheard worlds which come to open and expand before our eyes. At the artificial level we can invent only by means of successive simulations, while on the natural level we can only imagine what form successive illuminations will take. Here is the light of Grace which in Caravaggio comes to illuminate the cheat who plays with the Chaos of his life leading him to the metamorphosis-conversion. This is what happens to the mathematician who comes to change semantics by opening up to that onerous journey corresponding to the overcoming of himself as well as of his own vision of the world as an autonomous observer in action. The metamorphosis in the Other, and the opening up to a new world of thought and observation (together with the entry on the scene of new infinities) emerge necessarily to the extent of a radical transformation on the part of man, a transformation for which the right compass is not easily found. This is the challenge that life presents to us every day. Just think of the enormous load that as humans we carry on our shoulders: that load which every time requires the artist to work for the overcoming of his own Work as well as of what constitutes his inheritance as man and craftsman. Life grants no insurance in this sense; it cannot, in effect, give assurances to itself if it truly wants to succeed in ensuring the necessary renewal of its original creativity. Turing and Chaitin focus on the role played by the grid relative to pure software with reference to extroversion and disembodiment, the way is open to the identification of omega. This identification, however, turns out to be linked to a conception and previous petrifaction possessing a precise historical character. In this sense, therefore, omega has no absolute character: when, in effect, a real metamorphosis takes place, centered on the passage by the hero on his own shoulders, the coming into being of a new observation and, therefore, the very onset of new petrifaction come to enter the field. When the irruption occurs there is openness in depth, and unprecedented actors appear, albeit in the necessary context of an inheritance. Hence the proper sense of a natural evolution that can never come to be separated from the dialectic in place between Creativity and Meaning. The software leads, each time, to the opening from within of the hardware (with the birth of new intensities and the consequent outcrop of the God). The hardware leads, in turn, to the nesting in depth (in the swirls of meaning) of the software. The hardware opens on its abysses while also gifting incarnation to the point of surfacing as Nature. The software lurks deep within the sky of abstraction in successive increments of complexity (and its Methods) to outline the ever renewed contours of a kingdom of Culture. The pressure we will be under will not be of algorithms alone but also of meanings in action. The disembodiment must be pursued not only with a view to optimizing the evolutionary pressures on the table but also in view of a more ambitious goal: a complex system for living such as that represented by Ulysses–Marsyas appears to be the tool itself (first of all in overcoming the Pillars of Hercules), for an in-depth opening of its own hardware with the continuous birth of new meanings and with a continuous (but organic) remodulation of the ongoing evolutionary pressures. As Bergson states, complex living systems that act as autonomous agents come to enter the scene in function of ever new creativity, of the continuous realization of a renewed evolution. The way in which this happens is that which passes for the identification, each time, of the grid related to the martyrdom of St. Lawrence as masterfully illustrated by Titian. In the painting by Picasso "The flute of Pan" (Paris, Musée Picasso) the eigenvalues in action at the level of the score played by Pan are, in effect, the way to realize the trigger and the possible multiplication of the first cries of the Minotaur, that is to say of the first steps of the incarnation process. In this context, the eye of the mind, as well as the eye of Horus, son of Isis, appears, precisely, as one of the engines of natural evolution. But Horus is not only a name or a concept or an imagination that lives: he is, first of all, a universal Form (cf. Picasso's sculpture), that soul of itself and that guides and points to every possible vision. It is the eye that in Reflexivity becomes an eigenform to itself, a matrix of real invariance and autonomy. When we are faced with works that come to be worn by the Muse (Jacqueline wearing Picasso) through an ideal seam for files, we find ourselves faced with the offer of a particular inheritance from a piece of hardware that has come into being as an autonomous agent and that allows the new software (Marsyas as conceived) to come to light through the support offered by the cypher. Creativity along the path pursued by the Minotaur has turned into petrifaction, thus offering a legacy and while allowing the Goddess to conceive. Marsyas represents the new software that is born, but the hero is, in his turn, marked by a cypher, by a secret Rule that lives him in filigree: the robe relative to his sacrifice will be woven into the file of Reflexivity to the point of determining the giving of extroversion. At that moment the God will come to select opening to the new irruption (and new hardware).

Echo wearing the works is software that works with reference to the files of Reflexivity. A new Reflexivity will therefore only emerge after the extroversion and consequent selection. In order to go beyond omega we must, in effect, refer to a change in semantics, a change that must be sanctioned by the intervention of the selection operated by the God. The God who will come to select will be "discovered", therefore, along his subsequent surfacing as Nature: *Deus sive Natura.* This is why the selection by God is creative on a natural level. Hence the need not only not to forget the body (Varela and Carsetti), but also to come to realize the fact that only through a renewed incarnation will we have the opportunity to draw on a real overcoming of the ancient software. Thinking according to the *Factum* involves the very birth of new ways of thinking, and consequently of a new Self: here is the inventive activity in action. Marsyas who is conceived is also he who comes out of absence and marries (as Bacchus) Echo that ascends and wears the works of the Painter. Here is a coming to life through the fulfilment of death. Bacchus leaves a legacy managed by Pan, namely, the eigenvalues grid. Narcissus, in turn, leaves a legacy concerning the cypher related to the eigenforms. If for the embodiment the support is given by the eigenvalues that emerge on the surface of the body like a tattoo, when I work in the sky of abstraction the support will be given, instead, by the eigenforms that nest in the depth of the soul: body of the Minotaur and soul of the Painter. On the one hand, the grid- score, on the other, the cypher. When the Painter goes out of the absence it means that the Muse has come to wear the works according to the right "theory", intertwining them, that is to say, through the file related to the correct eigenforms, a file that is now given in the sky of Culture and that leads to new morphogenesis and no longer to invariance. Rather than drowning in the image, Marsyas regenerates himself through the fire in which he is to burn. His legacy is given by the eigenvalues, those eigenvalues that will come, in particular, to sustain the new irruption thus presiding over the consequent incarnation on the part of Minotaur. This is the metamorphosis in action. Files at the level of eigenforms and threads of meaning at the level of eigenvalues: cypher and score. The Muse who wears the works (and the Goddess who conceives) does so on the basis of the offer of the cypher, the cypher, that is to say, which allows the mirroring of the Goddess in Nature. Echo is the one who wears the works of Marsyas–Painter but is also the one who comes to conceive to the extent of being reflected in Nature, in the God who emerges as Nature.

To generate-conceive the Goddess must read herself in the file relating to Nature (and to the inheritance offered by Narcissus). She must read herself in the invariance, therefore, to give place to real (not illusory) morphogenesis. The Painter's works will have to be worn according to a file, the file related to a Nature *iuxta propria principia*. This will provide the passage necessary for the invariance, but, at the same time, also the passage from the reality of Nature to the sky of Culture. At the base of everything there is the relationship between Narcissus and Echo, as well as the relationship between the giving of the Road and the realization of the Assumption. Here is the peculiarity of the veil and carving at the level of the works: it is the carving that helps in the elevation of the soul: I will build my work like a cathedral (M. Proust) and it is the carving that allows for coming out of the absence on the part of the hero. The Goddess is mirrored and conceived to the extent that the Angel's offer is given. The cypher refers to the frame for fractals that subtends Nature (in terms of software). The Goddess who conceives, in this sense, is also the Muse–Echo adorned with the works of the Painter insofar as they are linked together on the basis of a hereditary file. The God, in turn, selects and breaks in as far as Marsyas' extroversion is realized together with the pyre concerning his thought with final identification of the score relative to the eigenvalues. We will, in reality, be able to ensure evolutionary change (albeit in respect of the conditions of invariance), only through our coming on our own shoulders (Skolem). Here is the importance of the research work carried out by Chaitin: to lay bare the mathematical structure of the brain as the principal organ of evolution means to place evolution itself before a possible new challenge, related to the realization and, at the same time, the overcoming of ourselves as real observers. But to this overcoming there will inevitably come to correspond an inheritance: our own inheritance. In this way we will come to be part of a History which is necessarily the History of Gods and humans together. It is in overcoming ourselves along the metamorphosis and in placing ourselves as creativity to ourselves, that a natural evolution can actually come about. Bacchus, in other words, will be able to carry out his work only along the extroversion of his brain, until he is selected by the God through the Method (Balthus). If Narcissus adds himself to the Temple, Marsyas comes to be added to himself by the God. Hence the appearance of the new irruption. When the Painter turns to the soul, the Muse is adorned with works along the watermark of a file. When the Minotaur returns to the body, Pan comes instead to play the score, thus coming to burn in the fire of the eigenvalues. Only if there is a watermark in conjunction with a file, can conception come about; just as long as the thread of meaning is present can the irruption come to be realized: eigenforms as a support for conception and eigenvalues as a support for extroversion and the consequent break-in. Once the hidden thread is represented by the file relating to the cypher-inheritance and once, on the other hand by the thread relating to the grid-inheritance: eigenforms versus eigenvalues. Once the inheritance is that of Narcissus who drowns in invariance and once that of Pan who burns in the stake, operating for morphogenesis through the change induced in Semantics.

References

1. Carsetti, A. (Ed.) (1984). Autopoiesi e teoria dei sistemi viventi. *La Nuova Critica, 64*.
2. Spencer Brown, G. (1969). *Laws of form*. London: Allen and Unwin.
3. Kauffman, L. H. (2018). Mathematical themes of Francisco Varela. *La Nuova Critica, 65–66*, [p. 72].
4. Fuchs, C. H. (2016). On participatory realism. arXiv:1601.04360v3 [quant-ph].
5. Andrade, J., & Varela, F. (1984). Self-reference and fixed points. *Acta Applic.Matem, 2*, 1–19.
6. Soto-Andrade, J., & Varela, F. (1990). On mental rotations and cortical activity patterns: A linear representation is still wanted. *Biological Cybernetics*, *64*, 221–223; Varela, F. (1990). Between turing and quantum mechanics there is a body to be found. *Behavioral and Brain Sciences (Commentary)*, *13*, 687–688
7. Varela, F., Thompson, E., & Rosch, E. (1991). *The embodied mind*. New York: MIT, [p. 8]
8. Varela, F., Thompson, E., & Rosch, E. (1991). *The embodied mind*. New York: MIT, [p. 9]
9. Merleau-Ponty, M. (1958). *Phenomenology of perception*. New York: Routledge, [pp. x–xi].
10. Dennett, D. C. (1993). Review of F. Varela, E. Thompson and E. Rosch. *The Embodied Mind*, *American Journal of Psychology, 106*, [p. 121–2].

11. Varela, F. (1995). The emergent self. In J. Brockman (Ed.), *The third culture*. New York: Simon and Schuster, [p. 154].
12. Varela, F. (1995). The emergent self. In J. Brockman (Ed.), *The third culture*. New York: Simon and Schuster, [p. 155].
13. Kauffman, L. (2009). Reflexivity and eigenforms. *Constructivist Foundations, 4*(3), [p. 123].
14. Kauffman, L. (2017). Eigenform and reflexivity. *Constructivist Foundations, 12*(3), [p. 250].
15. Kauffman, L. (2009). Reflexivity and eigenforms. *Constructivist Foundations, 4*(3), [p. 127].
16. Kauffman, L. (2009). Reflexivity and eigenforms. *Constructivist Foundations, 4*(3), [p. 121].
17. Carsetti, A. (Ed.) (2004). *Seeing, thinking and knowing. Meaning and self-organization in visual cognition and thought*. Dordrecht, The Netherlands: Kluwer Academic Publishers.
18. Benacerraf, P. (1965). What numbers could not be. *Philosophical Review, 74*, 47–73.
19. von Foerster, H. (1981). On constructing a reality. In *Observing systems* (pp. 288–309). Intersystems Publications; Kohonen, T. (1995). *Self-organizing maps*. New York: Springer.
20. Quinon, P., & Zdanowski, K. (2006). The intended model of arithmetic. An argument from tennenbaum's theorem. http://www.impan.pl/_kz/_les/PQKZTenn.pdf.
21. Carsetti, A. (1987). Teoria algoritmica dell'informazione e sistemi biologici. *La Nuova Critica, 3–4*, 37–66; Carsetti, A. (2000). Randomness, information and meaningful complexity: Some remarks about the emergence of biological structures. *La Nuova Critica, 36*, 47–109.
22. Carsetti, A. (2013). *Epistemic complexity and knowledge construction*. New York: Springer.
23. Petitot, J. (2008). *Neurogéometrie de la vision*. Paris: Ecole Polytechnique, [p. 397].
24. Kant, I. (2007). Critique of pure reason (2nd ed.). London: Mac Millan, [p. 153].
25. Chaitin, G. (1987). *Algorithmic information theory*. Cambridge, London.
26. Putnam, H. (1965). Trial and error predicate and the solution to a problem of Mostowski. *Journal of Symbolic Logic, 30*.
27. Chaitin, G., & Calude, C. (1999). Mathematics/randomness everywhere. *Nature, 400*, 3219–3220.
28. Martin-Delgado, M. A. (2011). *On quantum effects in a theory of biological evolution*. arXiv: 1109.0383v1 [p. 3].
29. Martin-Delgado, M.A. (2011). *On quantum effects in a theory of biological evolution*. arXiv: 1109.0383v1 [p. 4].
30. Chaitin, G. (2010). Metaphysics, metamathematics and metabiology. *APA, 10*(1), [p. 11].
31. Hernandez-Orozco, S., Kiani, N. A., & Zenil, H. (2018). Algorithmically probable mutations reproduce aspects of evolution, such as convergence rate, genetic memory and modularity. *Royal Society Open Science, 5*, 180399.
32. Chaitin, M. F. G. V., & Chaitin, G. (2017). A philosophical perspective on a metatheory of biological evolution. *Draft*, September 5, 2017, [p. 3].
33. Chaitin, M. F. G. V., & Chaitin, G. (2017). A philosophical perspective on a metatheory of biological evolution. *Draft*, September 5, 2017, [p. 8].
34. Chaitin, M. F. G. V., & Chaitin, G. (2017). A philosophical perspective on a metatheory of biological evolution. *Draft*, September 5, 2017, [p. 9].
35. Goedel, K. (1972). In Kurt Goedel: Collected works, I, II, III (S.L. Feferman et al. eds. 1986, 1990, 1995), Oxford: Oxford University Press, [pp. 271–272].
36. Carsetti, A. (2012). The emergence of meaning at the co-evolutionary level: An epistemological approach. *Applied Mathematics and Computation, 219*, 14–23.

Chapter 3
Non-standard Models and the "Construction" of Life

Abstract If we set ourselves from the point of view of a radical Constructivism, an effective semantic anchorage for an observer system such as the one, for example, represented by the non-trivial machine as imagined by H. von Foerster, can come to be identified only to the extent that the evolving system itself proves able to change the Semantics. This, however, will result in our being able to realize an expression of ourselves as autonomous beings, as subjects, in particular, capable of focusing on the same epistemological conditions relating to our autonomy. A creative autonomy that expresses itself above all in the observer's ability to govern the change taking place. Only the cognitive agent operating in these conditions will actually come to undergo the new embodiment. Here is the passage on one's shoulders to which T. Skolem refers, namely that continuous passage from the first to the second observer that marks the very course of natural evolution.

3.1 Non-standard Models and Skolem's Paradox

Let $\mathcal{L}$ be a fixed but arbitrary first-order language. By itself, $\mathcal{L}$ is meaningless: we have to introduce a basic semantic apparatus in order to endow $\mathcal{L}$-expressions with meaning.

1.1 Definition An $\mathcal{L}$-interpretation is an ordered triple Л consisting of the following components: (i) a non-empty set U called the domain or universe of Л. The members of U are called individuals of Л; (ii) a mapping that assigns to each function symbol $\mathbf{f}$ of $\mathcal{L}$ an operation $\mathbf{f}^{Л}$ on U, such that if $\mathbf{f}$ is a n-ary function symbol then $\mathbf{f}^{Л}$ is an n-ary operation on U; (iii) a mapping that assigns to each predicate symbol $\mathbf{P}$ of $\mathcal{L}$ a relation $\mathbf{P}^{Л}$ on U, such that if $\mathbf{P}$ is an n-ary predicate symbol then $\mathbf{P}^{Л}$ is an n-ary relation on U and such that if $\mathcal{L}$ has the equality symbol $=$ then $=^{Л}$ is the identity relation on U.

Starting from this definition it is easy to introduce the notion of $\mathcal{L}$-evaluation as well as the usual basic semantic definitions. For more technical details as well as for the adopted notations see Machover [1].

Now, let the formal object language $\mathcal{L}$ be the first-order language of arithmetic; namely the first-order language with equality $=$, whose extralogical symbols are: (i)

A. Carsetti, *Metabiology*, Studies in Applied Philosophy, Epistemology and Rational Ethics 50, https://doi.org/10.1007/978-3-030-32718-7_3

one individual constant, $\mathbf{0}$; (ii) one unary function symbol, $\mathbf{s}$; (iii) two binary function symbols, $+$ and $\times$.

Applying Definition 1.1 to this last language $\mathcal{L}$ we see that an $\mathcal{L}$-interpretation $\mathfrak{I}$ is determined in accordance with these essential elements: (i) a non-empty set U, the domain of $\mathfrak{I}$; (ii) an individual $\mathbf{0}^{\mathfrak{I}} \in U$, the individual denoted by $\mathbf{0}$ under the interpretation $\mathfrak{I}$; (iii) a unary operation $\mathbf{s}^{\mathfrak{I}}$ on U, the operation that interprets $\mathbf{s}$ under $\mathfrak{I}$; (iv) two binary operations $+^{\mathfrak{I}}$ and $\times^{\mathfrak{I}}$ on U, the operations that interpret $+$ and $\times$ respectively under $\mathfrak{I}$.

1.2 Definition The *standard* $\mathcal{L}$-interpretation $\mathfrak{N}$ is characterized as follows: (1) $\mathfrak{N}$ has as its domain the set N of natural numbers; (2) $\mathbf{0}^{\mathfrak{N}} = 0$; (3) $\mathbf{s}^{\mathfrak{N}} = \mathrm{s}$, the *successor* function; (4) $+^{\mathfrak{N}} = +$ and $\times^{\mathfrak{N}} = \times$.

1.3 Definition An $\mathcal{L}$-theory is a set $\mathbf{\Sigma} \subseteq \mathbf{\Phi_0}$ (where $\mathbf{\Phi_0}$ is the set of all $\mathcal{L}$-sentences); it is a set of $\mathcal{L}$-sentences closed under deducibility of $\mathcal{L}$-sentences.

The theory $\mathbf{\Omega}$ ($\mathbf{\Omega} =_{\mathrm{def}} \mathbf{Th}\mathfrak{N}$) consisting of all true sentences is called complete first-order arithmetic. $\mathbf{\Omega}$ can be considered as the whole truth about $\mathfrak{N}$ in $\mathcal{L}$. It is well known, however, that $\mathfrak{N}$ cannot be uniquely characterized in $\mathcal{L}$: even $\mathbf{\Omega}$ is not sufficient to single out $\mathfrak{N}$ because $\mathbf{\Omega}$ has, apart from $\mathfrak{N}$ itself, other models that are not isomorphic to $\mathfrak{N}$. Actually, the original version of the Löwenheim–Skolem theorem (1922) claimed that any theory which has an infinite model also has a model which is countably infinite. In 1933 (and 1934) Skolem restates his theorem as follows (in accordance with the translation proposed by Wang [2]).

1.4 Theorem (Skolem 1933) *There exists a system $\mathcal{N}^*$ of things, for which two operations $+$ and $\cdot$, and two relations $=$ and $<$ are defined, such that $\mathcal{N}^*$ is not isomorphic to the system $\mathcal{N}$ of natural numbers, but nevertheless, all sentences of* P *which are true of $\mathcal{N}$ are true of $\mathcal{N}^*$* [3].

P represents the underlying language in which the concepts of the elementary number theory can be formulated.

The crucial point of the proof consists in showing that $\mathcal{N}$ and $\mathcal{N}^*$ are not isomorphic. In an informal way, Skolem points out that: "La théorie récursive tout entière est valable dans $\mathcal{N}^*$. D'autre part, les modéles $\mathcal{N}$ et $\mathcal{N}^*$ ne peuvent être isomorphes; car alors il devrait exister une correspondence biunivoque conservant l'ordre, entre $\mathcal{N}$ et $\mathcal{N}^*$. Mais dans toute correspondence de ce genre les éléments $\mathcal{N}$ se représentent sur eux-mêmes, et l'élément $f(x) = x$ de $\mathcal{N}^*$ est > que toute constante, c'est-à-dire que tout élément de N. Dans le langage de la théorie des ensembles, $\mathcal{N}^*$ représente un type d'ordre beaucoup plus haut que N" [4]. In other words, Skolem by using diagonalization, forms a statement that holds in $\mathcal{N}^*$ but not in $\mathcal{N}$. This shows that $\mathcal{N}$ and $\mathcal{N}^*$ are not isomorphic: $\mathcal{N}^*$ is a proper elementary extension of $\mathcal{N}$. An important corollary follows from Theorem 1.4.

1.5 Corollary *No recursively enumerable axiom system using only the notation of* P *(i.e. using only concepts of elementary number theory) can determine uniquely the structure of the sequence of natural numbers.*

Corollary 1.5 can be also proved by Gödel's First Incompleteness Theorem, as Gödel himself notes in his review to a Skolem paper (Skolem 1933). "From this (Theorem 1.4) it follows that there is no axiom system employing only the notions mentioned at the outset (and therefore none at all employing only number-theoretic notions) that uniquely determines the structure of the sequence of natural numbers, a result that also follows without difficulty from the investigations of the reviewer in his 1931" [5]. In accordance with this line of thought it is possible to introduce the following Corollary as a consequence of Theorem 1.4 but proving it by applying the first Incompleteness Theorem.

1.6 Corollary PA *is not categorical.*

Instead of saying that PA cannot determine uniquely the structure of sequence of natural numbers, we can adopt the concept of categoricity to restate this result in a shorter way. Indeed, this will become even clearer when we look at Skolem's 1938 lecture in which he himself talks about the notion of categoricity and characterizes it in the very same way we use it today. As Skolem remarks: "On dit parfois qu'un champ est catégoriquement défini par certains axiomes, si deux modèles de ce champ, *M* et *M'* c'est-à-dire deux réalisations quelconques de ce champ, sont isomorphes par rapport à toutes les propriétés et relations dont il est question dans les axiomes, c'est-à-dire si *M* et *M'* peuvent être représentés univoquement l'un sur l'autre de telle façon que toutes ces propriétés et relations soient conservées dans la représentation" [6]. Skolem in his paper also considers the non-categoricity in connection with Gödel's Incompleteness Theorems in terms of existence of non-decidable theorems in PA which makes it a non-categorical theory: "... un résultat connu de K. Gödel, d'après lequel dans tout systéme formel ou dans toute théorie des ensembles qui embrasse l'arithmétique ordinaire, des théorèmes peuvent être formulées qui ne sont pas décidables. Si Σ est un théorème non décidable, il doit alors exister un modèle *M* pour lequel Σ est vrai ainsi qu'un modèle *M'* pour lequel Σ est faux, et alors *M* et *M'* sont sûrement non isomorphes c'est-à-dire qu'il n'existe pas de catégoricité" [7].

As is well known and as N. Di Giorgio points out appropriately [8], at the conceptual level, the relationship between Skolem and Gödel is a very intriguing relationship. At this proposal H. Wang expresses these enlightening considerations: "Since about 1950 I had been struck by the fact that all the pieces in Gödel's proof of the completeness of predicate logic had been available by 1929 in the work of Skolem ... supplemented by a simple observation of Herbrand's ... In my draft I explained this fact and said that Gödel had discovered the theorem independently and given it an attractive treatment" [9]. In a letter on 7 December 1967 Gödel casts some illuminant doubts on Wang's remark. "... It seems to me that, in some points, you don't represent matters quite correctly .... You say, in effect, that the completeness theorem is attributed to me only because of my attractive treatment. Perhaps it look this way, if the situation is viewed from the present state of logic by a superficial observer. The completeness theorem, mathematically, is indeed an almost trivial consequence of Skolem 1922. However, the fact is that, at that time, nobody (including Skolem himself) drew this conclusion (neither from Skolem 1922 nor, as I did, from similar considerations of his own). This blindness (or prejudice, or whatever you may call it) of logicians is indeed surprising. But I think the explanation is not hard to find. It lies in a widespread lack, at that time, of the required epistemological attitude toward metamathematics and toward nonfinitary reasoning" [10].

To sum up, we can note, from a general point of view and in accordance with contemporary terminology, that in 1934 Skolem affirms that there exists a non-standard model for $\mathbf{\Omega}$ (the whole truth about $\mathfrak{R}$ in $\mathcal{L}$) that is a model for $\mathbf{\Omega}$ that is not isomorphic to $\mathfrak{R}$. Moreover, there is such a model whose domain is denumerable. In this sense, $\mathbf{\Omega}$ fails to pin down $\mathfrak{R}$ uniquely (even up to isomorphism). As is well known, with respect to the original version of the Löwenheim–Skolem theorem (1922) as well as to its successive version (1934), a precise limitative result (called Skolem's Paradox) immediately arises when we notice that the standard axioms of set theory can themselves be formulated as (countable) collection of first order sentences. If these axioms have a model at all, then the Löwenheim–Skolem theorem ensures that they have a model with a countable domain. But, how can a countable model satisfy the first-order sentence which says that there are unaccountably many things? The original version of the Loewenheim–Skolem theorem appeared in a short paper full of important insights, but in Skolem's view the most important result contained in this paper is exactly represented by the afore mentioned Paradox: actually it is the first of the fundamental limitative results in logic. As Skolem writes in his concluding remarks, its importance is linked to the critique developed by the great scholar with respect to the reductionism in vogue. According to Skolem axiomatic set theory, as formalized for instance along the lines proposed by Zermelo and Fraenkel (ZF set theory), cannot be seen as a satisfactory ultimate foundation of mathematics.

Let us look more closely at the inner articulation of Skolem's Paradox always following (but in an very concise way) the lines of the presentation offered by M. Machover. Let we assume that ZF is formalized in a first-order language $\mathcal{L}$ with equality, whose only extralogical symbol is a binary predicate symbol ε. In the intended interpretation of $\mathcal{L}$, the variables range over all sets and ε is interpreted as denoting the relation ϵ of membership between sets. Let **ZF** be the formalized version of ZF. The postulates and theorems of ZF are expressed in **ZF** by $\mathcal{L}$-sentences. Moreover, let us assume that **ZF** is consistent: since the language $\mathcal{L}$ is denumerable it is easy to show that **ZF** has a model $\mathfrak{M}$ whose universe U is countable. What does the model $\mathfrak{M}$ consist of? In order to give an answer to this simple question we have to remark first of all that the individuals of the structure $\mathfrak{M}$ are what this very structure interprets as 'sets': the members of $\mathfrak{M}$ are $\mathfrak{M}$-*sets*. We know that all the theorems of **ZF** must be true in $\mathfrak{M}$. Among these theorems there is a sentence that says 'there exists an uncountable set'. Let us imagine an internal observer that inhabits $\mathfrak{M}$: the $\mathfrak{M}$-*sets* are the objects of his world. The observer can easily individuate a particular $\mathfrak{M}$-*set* h that instantiates this theorem. According to his point of view the members of h are really $\mathfrak{M}$-members of h, they are $\mathfrak{M}$-*sets* that bear the relation E to h, where E is the relation of $\mathfrak{M}$-*membership*. Hence an apparent contradiction arises: the real key of the Skolem's Paradox. The whole universe U of $\mathfrak{M}$ contains only denumerably many objects. How can there be uncountably many objects bearing the relation E to h? The resolution of the paradox from a syntactical point of view is simple: we have only to consider that many set-theoretical notions, such as countability, finiteness and so on are relative. A Л-*set h* may be uncountable in the sense of the structure Л even if when viewed from the outside *h* has only countably many Л-*members*.

In this sense, we can show that **ZF** has a model Л (with denumerable universe) such that the object $\omega^{Л}$ is non-standard: $\omega^{Л}$ also has Л-members that do not correspond to any natural number. If β is such a non-standard Л-member of $\omega^{Л}$ then β is a Л-finite-ordinal and results Л-finite. However, as seen from outside Л, β actually has infinitely many Л-members. Once again we are faced with the ultimate message concerning Skolem's Theorem: the structure $\mathfrak{N}$ of natural numbers cannot be characterized uniquely (up to isomorphism) in the first-order language of arithmetic. Actually, the set-theoretic characterizations of the system of natural numbers appear all relative. As Machover correctly remarks: "If mathematics—and in particular the arithmetic of natural numbers—is more than mere verbal discourse, then its reduction to axiomatic set theory somehow fails to do it full justice" [11]. From outside, we can see that the afore mentioned Л-system of natural numbers really contains infinite unnatural numbers.

3.2 What Does the Universe of a Non-standard Model Look like?

With respect to the discussion concerning the Löwenheim–Skolem theorem, many scholars have remarked that with respect to the philosophical evaluation of this theorem, we can also assume an epistemic point of view: we can consider, for instance, that we have access to a privileged class of models, the class, that is to say, concerning the intended models, and to the extent that we reveal ourselves able to describe it as completely as possible, we also have the possibility to come to know different kinds of models, called unintended, that undermine our knowledge of the intended models. We can for instance ask: "How many countable non-standard models are there and how they differ from the standard model of PA?". We know that there are $2^{\aleph_0}$ countable non-standard models of PA but on the basis of the following important Theorem 3.1 we also know that every countable non-standard model of PA is not recursive. Thus, by means of this theorem an explicit separation line is drawn between standard and non-standard models: namely the attempts to formulate arithmetical operations at the level of non-standard models will be severely restricted. However, we cannot forget that it follows from compactness theorem that the entire non-standard structure satisfies all the axioms of PA; i.e., that all arithmetical theorems are true in this structure. Actually, the theorem shows that if every finite fragment of a collection of statements has a model, then so does the entire (no matter how infinitely large) collection.

What does the universe of a non-standard model look like? According to Henkin, any countable non-standard model of arithmetic has order type $\omega + (\omega^* + \omega) \cdot \eta$, where ω is the order type of the standard natural numbers, ω^* is the dual order (an infinite decreasing sequence) and η is the order type of the rational numbers. In other words, a countable non-standard model begins with an infinite increasing sequence (the standard elements of the model). This is followed by a collection of "blocks", each of order type $\omega^* + \omega$, the order type of the integers. These blocks are in turn densely ordered with the order type of the rationals. The result follows fairly easily because it is easy to see that the non-standard numbers have to be dense and linearly ordered without endpoints, and the order type of the rationals is the only countable dense linear order without end points. Order-wise the model look like the natural numbers followed by densely many copies of the integers: $\mathbb{N} + \mathbb{Z}\mathbb{Q}$. Any countable model is isomorphic to one with domain $\mathbb{N}$. If the interpretations of 0, +, $\times$, and < in such a model are computable functions, we say it is a computable model.

The standard model $\mathbb{N}$ is computable, since the successor, addition, and multiplication functions and the less-than relation on $\mathbb{N}$ are computable. It is possible to define computable non-standard models of Q, but $\mathbb{N}$ is the only computable model of PA. From a more general point of view, we can naturally think of a countable non-standard model in terms of a non-standard structure $\mathbb{N} + \mathbb{Z}\mathbb{Q}$ that satisfies all the axioms of Peano arithmetic: i.e. that all arithmetical theorems are true in this structure.

We have just seen that the order type of the countable non-standard models is known. However, the arithmetical operations are much more sophisticated. Can we compute inside a non-standard model of arithmetic? Is there a non standard model of PA for which the operations + and $\times$ are computable? This question can be made precise in the following way. "With reference to $\mathbb{N} + \mathbb{Z}\mathbb{Q}$, we already know how to add a natural number n to a number (p, m) sitting in some block p $\in$ Q: we get (p, m + n). Is there a nonstandard model, where we would have an algorithm for figuring out (p, m) $\cdot$ n, or even better (p, m) + (q, n) and (p, m) $\cdot$ (q, n)?" As we shall see, Tennenbaum showed in 1959 that there is no countable non-standard model for which there is an algorithm to compute any of the above. Namely, Tennenbaum's theorem forbids arithmetic from having any countable non-standard model where either addition or multiplication is computable (in the sense of its countability, i.e. as interpreted in $\mathbb{N}$). As Gitman remarks: "The proof boils down to the fact the every countable non-standard model codes in some algorithmically undeterminable subset of the natural numbers and being able to compute (p, m) $\times$ n would give us an algorithm to determine which numbers are in that very complicated set" [12]. Actually, every nonstandard model of arithmetic contains non-algorithmic information. In other words, Tennenbaum theorem shows that for any countable non-standard model of Peano arithmetic there is no way to code the elements of the model as (standard) natural numbers such that either the addition or multiplication operation of the model is computable on the codes.

The arithmetical structure of a countable non-standard model is more complex than the structure of the rationals. In this sense, according to many scholars the existence of non-standard numbers in first-order arithmetic appears as a semantic obstacle for modelling our arithmetical skills. For instance Carnielli and Arujo argue that so far there is no adequate approach to overcome such a semantic obstacle, because we can also find out, and deal with, non-standard elements in Turing machines [13].

3.3 Tennenbaun Theorem: Every Non-standard Model of Arithmetic Contains Non-algorithmic Information

3.1 Theorem (Tennenbaum) *Let* $\mathcal{M}$*= (M, +, ·, 0, 1, <) be a countable model* of PA, *and not isomorphic to the standard model* $\mathcal{N}$*= (ℕ, +, ·, 0, 1, <). Then* $\mathcal{M}$ *is not recursive.*

We can also express the Theorem 3.1 as follows: Suppose that $\mathcal{M} \models$ PA is non-standard. Then neither $+^M$ nor $\times^M$ (i.e. the interpretations of the terms $+$ and $\times$ in $\mathcal{M}$) is recursive.

(1) We cannot hope to compute inside a non-standard model of arithmetic.
(2) Every non standard model of arithmetic contains non-algorithmic information.
(3) (ℕ, +, ·, 0, 1, <) is the unique computable model of arithmetic.

3.2 Theorem *There are continuum many countable non-isomorphic models of arithmetic.*

Proof Every subset of ℕ is coded in some countable model of arithmetic.

3.3 Theorem (Friedman 1973) *Every non-standard countable model of arithmetic is isomorphic to an initial segment of itself.*

3.4 Theorem *There are countable models of arithmetic with continuum many automorphisms.* (ℕ, +, ·, < , 0, 1) *has no automorphisms. A non-standard model of arithmetic can have indiscernible numbers satisfying the exact same first-order properties.*

Let us to see this up close. At a glance, the individuation of the Peano axioms intends to describe the "behaviour" of the natural numbers. However, thanks to Gödel's incompleteness theorem, today we know that these axioms can't completely capture the structure of the natural numbers. Hence the flourishing of a multiplicity of different 'models' of Peano arithmetic. These non-standard models can be countable or uncountable. Starting with any of these models you can define integers in the usual way and then rational numbers. So, there are lots of non-standard versions of the rational numbers. Any one of these will be a field: you can add, subtract, multiply and divide your non-standard rationals, in ways that obey all the usual basic rules.

On March 3, 2018 [14] J. D. Hamkins posted a very interesting theorem.

3.5 Theorem *Every model of* PA *of size at most continuum arises as a sub-semiring of the field of complex numbers* $\langle \mathbb{C}, +, \cdot \rangle$.

Proof Suppose that M is a model of PA of size at most continuum. Inside M, we may form M's version of the algebraic numbers $A = \bar{\mathbb{Q}}^M$, the field that M thinks is the algebraic closure of its version of the rationals. So A is an algebraically closed field of characteristic zero, which has an elementary extension to such a field of size continuum. Since the theory of algebraically closed fields of characteristic zero is categorical in all uncountable powers, it follows that A is isomorphic to a submodel of $\mathbb{C}$. Since M itself is isomorphic to a substructure of its rationals $\mathbb{Q}^M$, which sit inside A, it follows that M is isomorphic to a substructure of $\mathbb{C}$. QED.

As Baez remarks, the theorem states that: "*if your nonstandard model of the natural numbers is small enough, your field of nonstandard rational numbers can be found somewhere in the standard field of complex numbers!* (Emphasis original) Non- standard rationals are a subfield of the usual complex numbers: a subset that's closed under addition, subtraction, multiplication and division by things that aren't zero" [15].

As we have just seen, Tennenbaum's theorem forbids arithmetic from having any countable non-standard model where either addition or multiplication is computable (in the sense of its countability, i.e. as interpreted in $\mathbb{N}$). We have, however, also seen that we can say some things about what these countable non-standard models are like as ordered sets. Moreover, as Baez underlines, we know that what's good about algebraically closed fields of characteristic zero is the fact that: " they are uncountably categorical. In other words, any two models of the axioms for an algebraically closed field with the same cardinality must be isomorphic. (This is not true for the countable models: it's easy to find different countable algebraically closed fields of characteristic zero. They are not spooky and elusive)" [16]. Moreover, the theory of algebraically closed fields of characteristic zero is categorical in every uncountable cardinality, which means that there is up to isomorphism precisely one such field of each uncountable cardinality. This explains why any field of characteristic zero and size at most continuum is isomorphic to a subfield of $\mathbb{C}$.

3.6 Theorem (Hamkins) The *complex numbers* $\mathbb{C}$ *can be viewed as a non-standard version of the algebraic numbers* $\bar{\mathbb{Q}}^{M}$ *inside a nonstandard model M of* PA. *Indeed, for every uncountable algebraically closed field F of characteristic zero and every model of arithmetic* $M \models$ PA *of the same cardinality, the field F is isomorphic to the non-standard algebraic numbers* $\bar{\mathbb{Q}}^{M}$ *as M sees them.*

Theorem 3.6 show us how extensive are the mathematical waters in which we come to immerse ourselves once we enter the field of the non-standard models. But, is it possible to browse these mathematical waters? Is it possible to obtain new mathematical insights using the structure of non-standard models of arithmetic? Well, it is exactly using this structure that Kirby and Paris in the 1980s developed techniques to show that theoretic statements such ad Goodstein's theorem and the Paris-Harrington theorem cannot be proved from the Peano axioms [17]. As is well known, in mathematical logic the Paris-Harrington theorem states that a certain combinatorial principle in Ramsey theory, namely the strengthened finite Ramsey theorem, is true, but not provable in Peano arithmetic This was the first "natural" example of a true statement about the integers that could be stated in the language of arithmetic, but not proved in Peano arithmetic; it was already known that such statements existed by Goedel's first incompleteness theorem.

If a and b are natural numbers and $a < b$, use $[a, b]$ to denote the set $\{a, a+1, a+2, \ldots, b\}$. Paris and Harrington define a predicate PH (a, b) which says that the interval $[a, b]$ has a certain Ramsey-theoretic property. The assertion $\forall a\, \exists b$ PH (a, b) can be proven using the infinitary version Ramsey's theorem.

3.7 Theorem (Paris-Harrington) *Suppose a and b are non-standard elements of a model M of true arithmetic, and*

$$M \models \mathrm{PH}\,(a,\ b).$$

Then there is an initial segment I of M containing a but not b, such that

$$\mathrm{I} \models \mathrm{PA}.$$

|--------)---------|-----------)------------------|-----------------)

0 ω a I |= PA b *M*

3.8 Corollary PA *doesn't prove*

$$\forall a\ \exists b\ \mathrm{PH}\,(a,\ b)$$

Since Goedel's Incompleteness Theorems, for almost half a century logicians did not have examples of PA-unprovable statements that would not refer to diagonalisation or other logicians' tricks. The first PA-unprovable statements of 'mathematical' character (not referring to arithmetisation of syntax and provability) appeared in 1976 in the work of J. Paris (building upon joint work with L. Kirby) and led to the formulation of the Paris-Harrington Principle (denoted PH), a statement that is not provable in Peano Arithmetic. Many statements equivalent to PH have been studied: the Hercules-Hydra battle and termination of Goodstein sequences by L. Kirby and J. Paris etc. The statement can be expressed in the first-order language of arithmetic. It is easily provable in the second-order arithmetic, but is unprovable in first-order Peano arithmetic.

The theorem is very important at the logical and mathematical level, moreover it also possesses a major epistemological virtue: it shows us in a "natural" way as through the use of non-standard models it is possible to come to extend our awareness: we can, in particular, come to explore worlds never before known using, first of all, the theorems acquired through the exploration of the negative results.

3.4 The Structural Approach Versus the Wittgensteinian Approach

By Tennenbaum's Theorem we can realize that no construction of non-standard model can be finitarily given. It simply affirms that non-standard models do not belong to the world of constructive mathematics. The theorem gives rise to many philosophical discussions about non-standard models of PA. In particular, we can distinguish, in the first instance, two (among others) different approaches: a structuralist approach and a Wittgensteinian one. As we saw in the previous chapter, the structuralist point of view claims that the subject of mathematics is not made of numbers, diagrams etc., but rather mathematical structures. As P. Benacerraf remarks: "... numbers are not objects at all, because in giving the properties (that is, necessary and sufficient) of numbers you merely characterize an *abstract structure*—and the distinction lies in the fact that the "elements" of the structure have no properties other than those relating them to other "elements" of the same structure" [18]. In other words, the abstract (single) structure essentially appears as ongoing regulation.

The principal problem with respect to the structuralist point of view essentially concerns the possibility to actually rule out non-standard models of arithmetic. As a matter of fact, only in this way we could be able to adequately capture the single structure in question (up to isomorphism). An attempt in order to solve this difficult problem is that proposed by V. Halbach and L.Horsten through the introduction of a refined version of the original structuralist point of view called Computational Structuralism [16]. The general idea is to revisit Benacerraf in the light of Tennenbaum's Theorem. In particular, the two scholars make, first of all, essential reference to an informational notion of effective procedure as well as to a practical notion of computability in order not to presuppose number theory. Then, by appealing to Church's thesis they show that the theoretical and practical notions of computability are co-extensive. Finally, by resorting to Tennenbaum's Theorem they replace the theoretical notion of recursiveness utilized in the theorem by its practical version. In a nutshell, we are now essentially faced with algorithms and notations. Given that a notation system is a structure they, finally, can conclude as follows: "Intended [or standard] models are notation systems with recursive operations on them satisfying the Peano axioms" [19]. Thus, arithmetic lastly appears as essentially devoted to pure notations.

In order to introduce the second approach let us quote a passage from a paper by J. K. O'Regan, E. Myin and A. Noe which we introduced in the previous chapter: "... life is not something which is generated by some special organ in biological systems. Life is a *capacity* that living systems possess. An organism is alive when *it has the potential* to do certain things, like replicate, move, metabolize. etc." [20]. Always in the previous chapter we have seen how, in harmony with this general thesis by O'Regan et al. and contrary to structuralists, Quinon and Zdanowski [21] have come to point out that the basic property of natural numbers is the possibility to use them to count. They think that this property plays the decisive role in establishing whether a given model for arithmetic can be considered as being intended or not. As they remark, we actually learn what are natural numbers while learning to count. In their opinion, the intended model is a model that satisfies (and reflects) our intuitions concerning natural numbers. In this sense, they resort to the psychological version of Church's Thesis: "Any property that human can compute can be also computed by Turing machines" and link this thesis to the following two assumptions: computability of basic arithmetical operations and the principle of induction together with Tennenbaum's theorem. As a result, we finally have that the intended model for arithmetic is a recursive model with ω-type ordering.

At the level of this second approach we are dealing with a biological capacity that runs through our cognitive life and coagulates into a model that matches our intuitions. What to do, then, with the unintended models? Expunge them to recover the purity of a single structure as in the case of Computational Structuralism or open up again to Skolem's original perspective, the pursuit, namely, of always new formalisms if appropriate? If the latter appears, at present, the right way, the reference to Henkin is obliged: we have to achieve a compromise between the finitistic methods and the infinitistic ones and it is precisely on this rugged terrain that the general semantics can come to our rescue. Unfortunately, this is a path along which, in these last twelve years, few advances have been made if not episodically. This is a clear testimony of the difficulties that lie in wait. One thing, however, is clear (also based on the original insights provided by Henkin), the path of the research concerns in the first instance the determination of the size with respect to the order type. Here then, through Tennenbaum theorem first, the very necessity of a first, methodical inspection of the different order types and their inner articulation. But here is, also, the link that is to be operated, on the basis of the original insights envisaged by A. Robinson, with the kingdom of Non-standard Analysis, the same introduction of infinite numbers, the design of a real number line that takes into account the presence of infinitesimal etc.

The mathematical practice considered as a capacity that passes through us and that comes to live within us determining what we really manage to inspect as observers living at the natural level and permitting us to reach true existence, also manages, hand in hand, to reveal some of its mysterious recesses affording a glimpse of the deep dialectic underlying it. The fundamental problem is that we have to interrogate ourselves positing ourselves as oracles: "*Noli foras ire. In te ipsum redi*". The oracle and the inspirited truth can come to express themselves only in us and through us. We are within a path where the key is represented by the real possibility to change minds (and semantics) but in accordance with the transformation of our own fibers. Thus, we are no longer faced with intuitions concerning numbers as objects, their formation processes and their use. The numbers are natural because they arise from the progressive unfolding of specific primordial capacities. To the extent that these capacities embody according to specific cognitive processes they give rise to numbers, presiding over their formation in a natural way. Everything, however, is also linked to the identification of a Method: the "irruption" and the development of the original capacities is to the extent of the interrogation of the oracle and the exploration of the non-standard realm.

It is in this sense that, starting from the unfolding of a specific set of capacities (at the level of the original "magma"), counting appears closely related to the establishment of a recursive ordering. But if we look at this kind of process in a more accurate way, we can discover that counting is also related to self-reference and to the progressive unfolding of a tissue of eigenforms. Counting presupposes: (a) the full articulation of a set of capacities; (b) the actual existence of specific control procedures able, first of all, to manage recursions (in the light of an ongoing development of the software of meaning); (c) a process of mutual sharing determining the rise of specific forms of rational perception concerning the effective existence of real objects (the numbers) with respect to a set of observers; (d) the identification each time of an adequate set of meaning postulates. As we have seen, in the light of Set Theory and ZF, it is very easy to show that the construction of numbers is a recursive process. What is less obvious or less known is the role played by the fixed points at the level of that fundamental process that at the biological level undermine the construction of numbers in an operative way.

L. Lawvere in 1970 found a new way to prove a fundamental theorem by Georg Cantor which is at the basis of the development of Set Theory and the successive outlining of ZF. As is well known, Cantor's Theorem affirms that for every set S, there are more subsets of S than there are members of S. Today the right way to prove this theorem is first to prove Lawvere's fixed point theorem [22].

4.1 Theorem (Lawvere) *Suppose* $e: A \to B^A$ *is a surjective map. Then every map* $f: B \to B$ *has a fixed point, i.e.,* $x \in B$ *such that* $f(x) = x$.

Proof consider the map $s: A \to B$ defined by $s(x) = f(e(x)(x))$. Because e is onto, there is $y \in A$ such that $e(y) = s$. We have $e(y)(y) = s(y) = f(e(y)(y))$, therefore $e(y)(y)$ is a fixed point of f. QED.

4.2 Corollary (Cantor) *There is no onto map from a set A onto its powerset* $\mathcal{P}(A)$.

Proof $\mathcal{P}(A)$ is isomorphic to 2^A where $2 = \{0, 1\}$. In this sense, the subsets of A correspond to their characteristic functions $A \to \{0, 1\}$. Suppose $e: A \to 2^A$ is onto. But by Lawvere's Theorem every map $2 \to 2$ has a fixed point. On the contrary, this is not the case for the map $f(x) = 1 - x$. Therefore no such e can exist. QED.

We have just seen, by utilizing the Compactness theorem, that there is a countable non-standard model of arithmetic. On the other hand, we have just seen that, on the basis of a theorem by Henkin, the order type of any non-standard model of arithmetic is of the form $\omega + (\omega^* + \omega) \cdot \eta$. Nevertheless, it follows from compactness that this entire non-standard structure satisfies all the axioms of Peano arithmetic; i.e., that all the arithmetical theorems are true in this structure. In other words, we have that any countable non-standard model of arithmetic has order type $\mathbb{N} + \mathbb{Z}\mathbb{Q}$. Actually, up to isomorphism, the only countable dense linear order without endpoints is $\mathbb{Q}$.

However, we have also seen that if the order type of countable non-standard models is well known, the correlated arithmetical operations appear much more sophisticated. As a matter of fact, the arithmetical structure differs from $\omega + (\omega^* + \omega) \cdot \eta$. Tennenbaum's Theorem shows that for any countable non-standard model of Peano arithmetic there is no way to code the elements of the model as standard natural numbers such that either the addition or multiplication operation of the model is computable on the codes. In accordance with this line of research, we are faced, now, with some important theoretical developments. As a first example, let us introduce in view of a possible future deepening of the analysis the following theorem.

4.3 Theorem (Klaus Potthoff) *There is no non-standard model of arithmetic with order type* $\mathbb{N} + \mathbb{Z}\mathbb{R}$.

3.5 Computational Structuralism and Putnam's "Non-realist" Semantics

As W. Dean correctly remarks, Tennenbaum's Theorem can be understood to illustrate that: "although in classical mathematics we can demonstrate that non-standard models of arithmetic exist, the theorem intervenes to show that we can never hope to go beyond linguistic descriptions such as 'let $\mathcal{M}$ be a model of T_0'' so as to characterize the structure of $\mathcal{M}$ explicitly" [23]. Actually, given that $\mathcal{M}$ is countable we can characterize the substructure $\langle M, < M \rangle$ constructively up to isomorphism as the order type $\omega + (\omega * + \omega) \cdot \eta$.

However: "the fact that we have still gone on to develop a rich theory of such structures and their interrelationships is testament to the fact that the development of model theory often does not require us to fully extensionalize descriptions of models which we have introduced by such means" [24]. In fact, we can easily realize that our ability to refer to non-standard models must be understood as mediated by descriptions which are not only indefinite but which we know can never be made fully constructive. In other words, we are obliged to adopt a different understanding of 'model' inspired by Constructivism. Hence a possible confluence, at first, of the computationalist view with Putnam's 'nonrealist' semantics, with the attempt, that is to say, to identify the reference of an expression with its sense understood as an appropriate type of verification procedure. "'Objects' in constructive mathematics are *given through descriptions*. Those descriptions do not have to be mysteriously attached to those objects by some non natural process ... Rather the possibility of proving that a certain construction (the 'sense', so to speak, of the description of the model) has certain constructive properties is what is asserted and all that is asserted by saying the model 'exists'. In short, *reference is given through sense and sense is given through verification procedures and not truth conditions*" [25] (emphasis original). According to the theoretical perspective proposed here, let us remember that according to Benacerraf any set of objects with the ω-type ordering can be a model for arithmetic. However, next to this property we must also consider other important properties: actually, a basic feature of natural numbers is given by the fact that humans normally utilize them to count. Specifically we learn what natural numbers are while learning to count. But learning in the case of an autonomous agent (the Minotaur) is necessarily linked to the realization of an embodiment, an embodiment that, in turn, presupposes the encounter of the Minotaur with Ariadne as well as the full unfolding of his imagination. The agent must also take into account the use and the conditions of the exercise concerning his very self-identification: the properties relative to ω-ordering alone are not enough to identify the real exercise on the mat. In other words, the model is intended (and exists) when it adequately reflects our intuitions. "We learn what are natural numbers while learning to count. Consequently, we argue that an intended model for arithmetic should be such that one can perform basic arithmetical operations (addition and multiplication) on elements of this model (numbers from this model)" [26].

In this sense the model does not present itself simply as a construction characterized by certain properties: the model must also allow specific operations to be performed on its own elements and must have a privileged relationship with the use of certain abilities as operated by the autonomous agent at the level of the embodiment process at play. It is only if I prove myself capable of operating successfully on elements of the model that I come to understand: i.e. to exercise a specific skill such as that, for example, relative to counting. It is in this way that Ariadne can illuminate and that the construction of an I, in its turn, can be pursued.

We are far beyond Putnam: the model not only exists because it is identified through abstract structures, constructive properties and verification procedures, but also because it refers to the conditions proper to an actual embodiment of which, for instance, a specific learning process is an integral part. Thus, at the level of the intended model for arithmetic we have the convergence of recursivity, first-order induction and ω-type ordering (Tennenbaum theorem): the intended model necessarily takes shape in reference to specific recursive processes, to what is, from a general point of view, the landscape of Reflexivity. A convergence, in any case, that takes place in the context of the detachment operated by the Minotaur. What happens, however, when we enter the arena of metamorphosis in all its breadth? When, that is to say, we take into account the entire journey of the Minotaur. In such a case, as Picasso clearly shows in the painting "The flute of Pan", a further element enters the scene: the score relating to the inheritance of Pan, i.e. the original set of the eigenvalues on the carpet. We are now in the realm of non-standard models where the reference to the ordering is to vary as shown by Henkin in 1950. We will no longer only be faced with ω-ordering but, for instance, also with the order type $\omega + (\omega * + \omega) \cdot \eta$ and so on. The imagination at work at the level of the embodiment (as, indeed, shown by Picasso in his painting) is guided by eigenvalues and not by eigenforms. Let's imagine now to recover, in the footsteps of Carsetti (1990), a suitable model for a given process of metamorphosis and self-organization [27]. Having to refer, initially, to a set of eigenvalues, it will be necessary to refer not only to recursive processes and standard models but also to both non-standard models and simulation and invention procedures. Hence the entry into the scene of a new theoretical perspective: the perspective related to set-theoretic Relativism. Now, we must adopt, as Skolem does, a different understanding of 'model' inspired by Constructivism and set-theoretic Relativism. The object-construction to which Putnam refers is now replaced by a process of self-organization, by the very decline of a metamorphosis process such as that so well illustrated by Picasso or Ovid. We are faced with a dialectic at play between imagination and invention and not only with the presence of specific relations between objects. In the light of this new perspective, we should maintain that to find out which algorithms really correspond to the references relative to some specific constructive operations it should mean for the autonomous agent that undergoes the metamorphosis to be able to make Nature speak ("constructing" and interrogating in the right way the oracle as a new Oedipus) in order to come to feel the solution of the problem in its coming to flow at the level of his own veins. In this sense, only an effective renewed embodiment can, therefore, tell us what the algorithms in question should be. It is the new life and with it the new arising mathematics that will come to condition the self-organizing fibers of the Minotaur along the course of his own evolution starting from the actual giving of the irruption as it arises from the sacrifice of Marsyas. A life, in particular, that will, then, extend itself along the profiles of a new invariance (up to the self-organizing of a Road, but in the silence, a silence interrupted only by the flowing and fading away of the sound related to the stiletto heels of Echo). The reference for an autonomous self-organizing agent is given by the achievement and verification (but on his own flesh) of his autonomy, the autonomy proper to an agent that manages to handle the algorithmic schemes at work in accordance with his inner transformation thus resulting able to prove that he exists to the extent that he places himself at the root of the fulfilment of metamorphosis.

The new autonomous agent who will thus be born will therefore be able to look at the ancient remains of the first observer thus realizing, as Skolem himself states, that many ancient figures which inhabited the theoretical universe of the first observer (such as the ancient infinities) no longer show themselves in accordance with their original characteristics (i.e. as true infinities) with respect to the new arising horizon (the horizon relative to the new observation that is born). We are, in effect, faced with a new embodiment and the conditions relating to the model will now undergo a radical change.

If we set ourselves from the point of view of a radical Constructivism, an effective semantic anchorage for an observer system such as the one, for example, represented by the non-trivial machine as imagined by H. von Foerster, can come to be identified only to the extent that the evolving system itself proves able to change the semantics. This, however, will result in our being able to realize an expression of ourselves as autonomous beings, as subjects, in particular, capable of focusing on the same epistemological conditions relating to our autonomy. A creative autonomy that expresses itself above all in the observer's ability to govern the change taking place. Only the Minotaur operating in accordance with these conditions will actually come to undergo the new embodiment. Here is the passage on one's shoulders to which Skolem refers, namely that continuous passage from the first to the second observer that marks the very course of natural evolution.

We will then be able to place ourselves as witnesses (but at the level of the new embodiment) of what in the past has been the ability on the part of the first observer to govern his own growth process. Here is the flourishing of an intentional logic based on the ineliminable relationship with the Other. At the outset there is no ability to count, in fact the eyes of the Minotaur as painted by Picasso are not open from the beginning: they come to open as genuine eyes only to the extent of the construction in progress of those structures of imagination that allow the correct articulation of the schemes and, therefore, the same birth in the round, but by *bricolage*, of the activity of counting. Biological and cognitive activity is always in reference to the evolution at work and the construction of a Temple intersected by perceptual acts (within the framework of the ongoing dialectic between incompressibility and meaning).

The reference for an autonomous self-organizing agent is given by the achievement and verification (but on his own flesh) of his autonomy, the autonomy proper to an agent that manages to handle the algorithmic schemes at work in accordance with his own transformation thus resulting able to prove that he exists to the extent that he places himself at the root of the construction of the properties that identify his very creativity: true existence is given by creativity at work (but in the agent's awareness of this same creativity).

As we will see on the following pages and as Tennenbaum teaches the transition to the non-standard realm, the reduction that is given to the many sorted case and so on, makes it impossible for the Minotaur as an autonomous agent to close himself from the beginning in a framework studded with pure eigenforms. This will only happen at the moment of the petrifaction and the fulfilment of the process related to the ongoing self-organization. Until the break-in from the bottom persists, the evolutionary functions will grow up on themselves and cannot be used as witnesses of themselves except in the context of a specific *bricolage* in agreement to which they will play a circumscribed role. As we have just said, at the outset there is no preformed ability to count: the Minotaur is blind. The invariant tissue of eigenforms represents a conquest and not a starting point.

Everything must be brought back to the intuition of Skolem, concerning the passage on his shoulders by the first observer. We move from number systems to semantic and self-organizing databases. These databases can only be inhabited by chaos: hence a mathematics of Nature that presents itself as a mathematics of fractals. Here is the importance of the appropriate reference to many sorted algebras. Hence the need for the transition from the first order level to the second order level. At the level of Nature, the laws of Quantum Mechanics, the laws of chaos, the laws that lead to the construction of the mind and to the actions of the observer etc. apply. The basic problem is that of sharing information but at an intentional level. Hence a Nature growing on itself, a Nature that appears to be populated by a multiplicity of different I as related to specific minds but in connection with the reality of their brains. The transition to the non-standard realm represents the fundamental element of this reality in motion that cannot, in any case, be separated from the actions of the observer. This naturally has a price: that of metamorphosis. A price that is linked each time to the abandonment of invariance, to the "construction" of the *experimentum crucis* and so on.

When the Minotaur places himself as a model to himself, closing himself in Reflexivity, this presupposes that the legacy of Marsyas has come expressing its leadership at the level of the surfacing in progress. Now, they are the eigenvalues on the carpet that preside over the decline of the eigenforms. Here, then, is a complex dialectic that sees the presence of a specific inheritance as well as the intervention of adequate orderings in accordance with Ariadne's assumption. It is in this framework that one has access to the activities of counting (as well as to other activities specific to culture). The reference to the guide offered by Pan is permanent and it will also be in reference to it that it will therefore be possible to change semantics thus accessing, one more time, unheard universes as well as forms of computation never before realized. A new observer will be born but on the basis of coming on his shoulders by the first observer. Ancient infinities will no longer reveal themselves as such and new horizons will present themselves in the eyes of a Minotaur who will come to be born from the burning of renewed intensities. Then, it will be the second observer who will come to see the inert assembly relative to the ancient remains of the first observer as a borderline case that fits, given certain conditions, within what is now his perception, a perception that reveals itself to a wider extent as living perception: the perception proper to a cognitive being. Here is a new perception concerning growth and metamorphosis as well as the ascent to the sky of Echo and the subsequent channeling of the oracle. The verification by the Minotaur of his own sense is given, therefore, in the metamorphosis: here is the shepherd who looks at his own memory but adding himself to the Temple and opening to new conception. The passage for Tennenbaum and for radical Constructivism is, in this sense, necessary. When we enter the new world a new life and a new way of seeing will come to manifest. The verification concerning the operations and the programs used by the first observer must now come to refer to the new worlds that have come to be released following the metamorphosis.

From a general point of view, let us underline that, at the level of Henkin semantics, the problem lies not only in the identification of new orderings, but also in the opening up to a wider range of capacities capable of triggering a more pervasive canalization as related to a targeted introduction of new elements and relationships. Hence the need to refer each time to a new "assumption" with respect to the software of meaning at play as well as to the identification of a new game of life, a game characterized by unheard roles and rules. When this happens we shall necessarily be faced with the emergence of a new stage at the level of the many-sorted case, and we will be forced to resort to a methodical and targeted elimination of relations as well as to the introduction of restricted quantifiers, in view of giving rise not only to the ideal structure of a Temple (the actual ordering concerning the model) but also to the carving concerning it, the same totality of its reliefs, in order to further forge (and explore) the related meaning. Thus the Temple will render itself as a cathedral and a story with endless semantic intersections. When, on the basis of the ongoing control on behalf of the new software of meaning at stake and the effected irruption, a specific channeling process comes to be properly accomplished, then we shall be faced with the effective realization of an embodiment as connected to the appearance of new eigenforms. Hence new operations on and through numbers, the rising of a new mathematical language operating on a natural level. One thing is, therefore, the single structure itself in isolation while quite another is the ongoing synthesis concerning the three fundamental dimensions characterizing the development of a living (and cognitive) system: capacity (with unfolding), control (with interrogation) and sharing (with communication). The canalization, in fact, is also dependent on the "assumption" made and the pursuit of the path of the metamorphosis (along the embodiment process), of that path which, in its turn, is in proportion to the progressive emergence of an "observer" starting from the set of the original intensities at play. One thing, in other words, is an observation activity which does not allow the model to open up, thus preventing the possibility of a regulation of regulation and another thing is a simulation activity able to establish a new link with creativity in action. A new synthesis will then come about but on the basis of the "assumption" which will have been realized again; new eigenforms therefore will be born but in an unpredictable way, and with them new systems of numbers. Hence the emergence of new constraints with respect to original information with consequent birth of new and unprecedented capacities.

In this way, a specific tuning, which will always prove different according to the code and the invariants at play, will come to determine each time our existence as observers and as craftsmen. In the "arena" of such tuning, for example, the systems of numbers acquire concreteness and grow on themselves on the basis of the definition of ever-renewed eigenforms and metaeigenforms, but in dependence of the realized "assumption". Hence the central role played by Henkin semantics and the software of meaning at the level of the information process characterizing the biological reality. Hence, again, the possibility of our re-appropriation of ourselves, of the realization by means of our own hands of a precise carving of the ongoing "inspiration" flow as well as the same opportunity for us to become partakers of this particular process. We can rely on and work with numbers as we can refer to specific eigenforms operating in the field of reference, but this is dependent upon the control exercised by the model (and the related software) at the level of the original capacities that open to the world. When the control is carried out properly, the interrogation may, in turn, come into play. Hence the *experimentum crucis* and the exploration of the Underworld. In this sense, for instance, the inspiration goes through my bandaged head (Capogrossi, "*Il Vestibolo*", 1930) or through the work of Morandi and the glimmers of light that flow into one another and return the memory and the profound reality of the real objects, the very essence of things: that essence that allows the preservation and the resurrection of the body (but along the passage in the tunnels of the Underworld). When the non-standard models come about precise changes occur at the level of language in action but according to a deeply-carved path. There is no longer architecture, but dismemberment, an absence that comes to live in our interiority and devours inside our certainties but that open to a new core. Starting from the tunnels crossed from the infinitistic Methods a new content of intentional information will come to be recovered but in the presence of a renewed "assumption" of the actual meaning.

Here is the importance of Skolem and the lesson of non-standard mathematics. Climbing up on our own shoulders, working outward and opening ourselves to the selective intervention of the Method, we set the conditions for the emergence of new observation conditions. The new observer will therefore be able to compute with respect to a finitary realm specific "realities" (sequences) that were previously considered as infinite. But this will only be possible in proportion to the stumbling block in his own body along the run on the beach by the new observer that is born. Here is an observer who comes to the flesh but in the presence of the Pan flute and the eigenvalues on the carpet. (Cf, Picasso: "*La course*", Musée Picasso, Paris). But it will be the emergence of new observation conditions that will then determine the new modulations of the revisable thought as it will be given at the level of the craftsman's action and the new path in abstraction. From the new capacities-intensities to the new conception based on the offer by the Angel and the triumph of Reflexivity. Now new number systems can be identified but to the extent of the *experimentum* performed and the related death in the flesh. The reality of such systems will be given by their coming to act as yeast for new possible observation conditions. Here is the sense of the first observer's passing on his shoulders and here is the advent of that reality that identifies the *Factum*: *Verum et Factum convertuntur*. There can be no perception without thought and vice versa. Then, it will be those particular bodies that will have come to realize themselves in the stumbling block that will evolve in the conditions of their emergence and it will be in reference to them that the selective pressures will come to loom from time to time. To open to the new *Sylva* it is not sufficient to simply disregard the body as Chaitin does: it is necessary to resort to the right *experimentum* and it is necessary to govern the new irruption. It will also be necessary for the Minotaur to emerge starting from the detachment and the subsequent determination of the observation conditions (with consequent and progressive opening of his eyes, an opening this last to which the conquest of his autonomy by the Minotaur will come to correspond). Hence the necessary petrifaction and the new Assumption. In other words, it is not enough to immerse the possible game of selective pressures in the abstract space of a given software, thus cancelling the embodiment. Actually, the disembodiment is always relative to the construction of a new possible asset with respect to the game of life with the consequent and continuous appearance of new rules and unheard objectives. Hence new perceptual acts and new Nature, new DNA in action and new emotions. Here is a Minotaur who as shepherd comes to uncover the sepulcher and together with it the code and the memory but new. However, without guidance and control, without the mentor (Virgil the classic in action) there is no hope of mastering the *Sylva*. Hence the need for an innovative relationship with biological complexity. It is the model linked to Ariadne (as related to Virgil the classic) that must accompany the subject-hero in the Underworld.

3.6 Meaning, Conceptual Complexity and Intentionality

In rendering himself visible at the selection level and burning inside in the fire of morphogenesis, Apollo constitutes the *locus* of intensities, a place always renovated which opens to the process of new incarnation and new vision. The observer who realizes the detachment within himself will, then, feel in his veins, reborn to the life of the flesh, the flowing of the *Sylva*, and will feel in his own breast the activation of a new categorial as the new embodiment comes into being. In this sense the Mountain (the Cezanne Mountain) once made visible at the selection level will come to constitute the original reference point with respect to the new observer as he emerges along the path of incarnation: the path, that is to say, subsequent to the fire and selection set in action by the God. Here is the pyre of the thought into which the painting will come to insinuate itself and ultimately spark into life. The painter will then come to think of himself as burning on account of the Work deployed, and will experience rebirth, as a new Lazarus, at the level of the painting which will succeed in sparking into life at the level of the fire. Should this come about it could then be followed, (after successful incarnation), by detachment, and the newborn observer (on a par with other observers able to complete the same path) will reveal himself as the witness to the power of the God (and of the Mountain), in agreement with an eye, that of the mind received and set in stone. Within such a framework Ariadne's nourishment of Bacchus-Marsyas represents one of the two essential points in the process. Marsyas will succeed in playing the flageolet of possibilities insofar as he has fed on meaning; he will be able to go off exploring the reality of possible worlds until he draws the instrument (the spinet) with which he will self-identify, and which will allow him to tune the new beats of the world, inventing new chords for what has to be said and thought, ensuring that the language of which he is the Son may become thought. In this sense he could hardly fail to be selected by Apollo. The light which nurtures him ultimately presides over the extroversion of his cortex. In this way a language becomes thought, each time born anew through the fire of irruption where in its turn a thought had surrendered to language in the waters of conception. It is, however, the flayed Silenus who leaves the trace of the Ideas on the bloody ground. If the cipher allows conception, the trace allows the new irruption, constituting the necessary support for the fire of thought to burn once more, (and for a new universe of the categorial to come to exist). It will be from the new pyre and the trace of the Ideas asserting itself imperiously that a new path of incarnation will ensue, with the concomitant new detachment. It will now be the Imagination which is pressed into service and called to the work. The conquest awaiting the hero born from the transmutation will be the fruit of the hero's vision in the truth, but equally in agreement with the emergent Reflexivity.

The hero offers, *in primis*, himself as catalyst and actual canalization for the nesting by *bricolage* of the hardware. As craftsman and agent, he will once burn in the fire, giving way to the emergence of new thought (which will in turn change into a new categorial and new intensities), while as observer he will once emerge as canalization for a new surfacing of the software as he manages to fix on his own image in the waters. In this sense, if Ariadne illuminates by way of specific orderings, Apollo shapes and enforms by way of Ideas. If the Form is determined (and illuminates), the Function, conversely, is triggered (and generates). The Ideas prepare the irruption: the orderings the conception. Conception is *via* systems and orderings: irruption *via* Ideas. Marsyas invents, first of all, by means of complex numbers, only after he will use infinitesimals and non-standard numbers. Once he has burnt in the fire of new creativity he, the newly-born Theseus, will be able to explore the steps towards a new incarnation, starting from the stirrings of his brain which will open to the message-impressions of the Mountain. More particularly, he will come to see with new eyes, through new imagination and new mind, and as such will be illuminated by the Goddess by means of rules: rules which freeze rather than burn within themselves. If the Mountain itself burns, the lake of reflexion and self-referentiality takes as reference-point an observer enclosed within cyclical and invariant laws.

The observer is nourished by the sacrifice of the Painter and by the Painter's representation of the trajectory of his sacrifice, the sacrifice which allows the creativity burning inside the Mountain to find itself again by operating its recovery. When this occurs the external observer is in actual fact supported by an unprecedented set of intensities as well as by a renewed categorial. The way of abstraction once taken, the road is now open to a renewed process of incarnation. The intuition by Skolem concerning the dialectic between the two observers now become relevant in that it is precisely on account of this dialectic that we must each time return to the continuous mediation of Marsyas. It is only Silenus' passage through the Underworld which can allow the second observer to see beyond: i. e. to climb on the shoulders of the first observer through reference to the procedures of undecidability and the process of incarnation. The Painter leads me by the hand to the emergence of a new categorial, whence the possibility of new vision. I shall be able to witness a new, true reality having passed through the eyes of the resurrected Lazarus, having learnt from selection the correct Method in view of fulfilling my resurrection and the passage to the new vision, since I shall have awoken in the flesh of the new observer who rises and observes the ancient remains in the precise awareness that they are illusory. From this derives the discovery of new (and true) infinities which therefore can only reveal themselves, in their turn, as relative, as a consequence of a renewed conception and new Assumption. Balthus must posit himself as the painter–creator: he who, like Paolo Veronese, permits the resurrection of the son of the widow Nain (and of Lazarus). When this occurs he rises again within the praxis of painting: at the same time, however, he opens to the possibility that other observers (and he himself, as the observer reborn to the vision on emerging from the *Sylva*), will turn their gaze to the very act of resurrection, albeit with new eyes. These eyes are those of a lowly assistant aware of his status but able to question himself and consider his own history in such a way that the other observers can question him and penetrate deep within him. Observation, awareness, and self-organization: this is the cipher of the painting "La Rue" by Balthus, a painting which apparently is arrested in the immobility of an interior Time which will burn in the awareness of the God (of the Other) through the Painter's adjunction to the original creativity after Lazarus' resurrection. Here is a creator who will then self-detach after being added, though within the Other and in necessary communication with the Other as the necessary transmutation (metamorphosis) comes about. The Time is always that of the resurrection (and rediscovery) and as such is eternal; it must however be renewed each time according to the praxis of painting and the *experimentum crucis* as experimented by the painter at the level of successful extroversion. Without Marsyas the new observer is unable to enter the field: the renewal must necessarily be part of the actual forming of the filter relative to his consciousness.

Ariadne provides semantic information by opening to invention. Marsyas computes and submits to selection (the Method). Apollo burns within himself, while the observer detaches himself: hence imagination at work and the enlightening on the part of the Goddess. The Mountain offering the data is Ariadne, though the selection is then activated by Apollo. The Mountain is Ariadne who nourishes, and simultaneously Apollo, or selection which jolts and shakes. The creator creates-invents the painting and is able finally to grasp the Mountain insofar as it selects him (during the achieved extroversion). The artificial painting created by the Silenus' hand is necessary if the Mountain is to recover its creativity. Without Ideas and without Philemon and Baucis no resurrection is possible (or life at the level of the Mountain), but only a waste land. The painting (the painter's artificial work) which finds life once more in the fire ensures the recovery of creativity on the part of the God and the adjunction in the soul of the creator. Besides the fire and the *Sylva* there then follows the new incarnation and the activity of vision on Narcissus's part (no longer selected by the God but illuminated by the Goddess). The latter nourishes, the former selects; the God generates and categorizes, the Goddess illuminates. A film montage takes place which externalizes the trajectory of a brain which thinks of the Mountain, succeeds in representing it in a painting, and actually offers a Still Life which is each time able to reactivate and rebind the activity of nourishing on the part of the Goddess. If, however, this artificial Work is responsible for rebinding, it is also true that it rediscovers its own life in the flames: here is the secret of a true Still Life. The extroversion of the cortex by means of the Work and selective intervention on the part of the God allows the adjunction of the hero and the fulfillment of recovery. The artificial painting can now itself animate the Mountain and make it visible to the observer born of the effected transmutation. Through the Still Life the Painter renews his invention to the extent to which it ensures new nourishment on the part of Ariadne. This in its turn occasions new selection, and thus possible new incarnation: the birth of a new Nymph, but in parallel with the Assumption of the Goddess. The externalization regards the artificial painting which comes to life, but for the Other. The artificial painting assumes life at the level of thought-consuming fire, and it is in this coming to life that the true Mountain also comes to earns its existence, as the fire progressively reveals its profound logic. The creator has grasped the secret, though necessarily in the transmutation which will lead him to the new observation in the flesh (and in the detachment) as well as to the final offering of the flower.

The Goddess conceives and nurtures Marsyas (the simulation language that is born). The conception takes place at the end of the way of incarnation. The Goddess who has enlightened the final journey of the Minotaur by means of specially dedicated orderings, thereby permitting the activities of vision and reflexion to come about, can finally bring about the nurturing of the Silenus at the moment in which she is able to conceive through the offering of the flower. It will, then, be the true Mountain which selects the developing of the child Bacchus, and is the same Mountain (as Apollo in action) which through the Method will come to select the Silenus, who is ready to offer himself as creator, having completed the building of the spinet. Marsyas plays but does not create; he simulates and calculates, but artificially. However, by offering himself as pure artifice he is added to the Mountain in the awakening of the great fire of life. When this happens the new categorial is again active, and imagination and detachment are once more at work. Now a Son not of simulation but of the flesh can be resurrected. At the moment of resurrection, his first glance will be towards the ancient remains of the Silenus: something he can do since his path is illuminated by the Goddess by means of the orderings. Hence the hero, dressed in flesh, who rises and questions himself. The Minotaur who is born will succeed in opening the eyes of flesh as a thought is turning into language. This results in imagination at work as well as in specific figures of judgment. Hence the first articulation of a mind. The Minotaur who is born has no need of information as nutriment, being already thought: a thought which must be channeled through the imagination. The result is a mind expressing itself as language and as a linguistic system, thus allowing the ongoing resurrection to express itself in accordance with the surfacing of the correct software. The Minotaur is precisely the Demiurge of this surfacing, and only through dependency on this can the hero manage to open his eyes in the truth. It is however Ariadne who enshrines the entire process through illumination, or rather, through enumeration by means of orderings. When Ariadne nurtures and illuminates giving shape to a universe of imagination, this is then able to give itself progressively to life and generativity at the level of a Nature which ends by revealing itself as *iuxta propria principia*. On the one hand, here is a mind which acquires form in the radiance of flesh; on the other, a brain shaping itself at an artificial level in the darkness of abstraction. Eigenforms versus eigenvalues: reflexion within invariance in one case, and burning in metamorphosis in another. Death and life, cyclicality and resurrection. As creator, Cézanne is initially nourished by Ariadne, but once added to himself (in resurrection) by the God, he opens to transmutation in the observer operating the detachment. At that moment Ariadne will again be able to select the Minotaur who will have succeeded in articulating the schemes concerned with the imagination at work. Once Ariadne nourishes and once enshrines selecting by enumeration. Equally Apollo (the Mountain) who selects by elimination is also he who, as *Sylva* in action, nourishes the Minotaur operating the detachment both with his Method and the categorial. Narcissus operates the fixation with regard to Ariadne, the Painter the metamorphosis, with regard to Apollo (through extroversion). By offering himself as instrumentation the Painter determines both the soul's renewal and the change of the thought process through Ideas. The result is the Mountain that "speaks" a new thought, and an observer who realizes a new Physics and a new mind within himself. Nature self-expresses through denotation, the Work through connotation. Nature speaks a language which is mathematical, while the Work spreads like a hurricane of thought at a physical level. It is at the level of the fire that new thoughts, a new categorial, and new intensities are triggered according to Ideas. The observer still to be born will refer to a new Physics and a new conceptual framework, the framework. With great precision, through which he will proceed to observe the world.

To sum up, when, departing from initial incompressibility, a type of channeling is fully realized at the mental level through a specific metamorphosis and in the presence of an anchorage *via* forms supported by the software of meaning, then we are faced with the concrete ostension of a model: the model, for example, which reflects (i.e. satisfies) in itself our intuitions concerning natural numbers in accordance with a specific mathematical practice. Coming to perceive these numbers actually means counting and articulating them in keeping with a process of specific learning, and allowing room for a set of capacities (for the most part mysterious) which lives (and spreads) within us. A model—in the specific case, of the system of natural numbers—should precisely allow the correct performing of operations (amounts/sums etc.) on elements of the model itself. At the end of a long journey we thus close ourselves within a world of "safe" operations which for us represents the intended model in which we reflect and fix ourselves and through which we fix (and pave) the reality of our cognitive life, thereby helping to render the world in which we then come to act a mathematically defined world, inscribed as such within itself. This world reveals itself, however, as mathematical (Nature inscribed in mathematical characters in which we operate as observers), only insofar as it is traversed by life. The model sets us against our achieved capacity to count and produce numerical sums etc., providing the elements, operations, and mathematical framework to perform operations within a computational context. This satisfies our intuitions or reflects in itself our ability to use numbers to count; it permits us to reveal ourselves to ourselves, as subjects able to act mathematically according to conditions which are semantically correct. In the case of 60s Minimalism, for example, the mathematical structure as everted in an image (notably the image related to the Fibonacci numbers as they appear in a celebrated painting by D. Judd) constitutes the means whereby we reflect (but as artists and at the conceptual level) in a work of Art the mysterious order of our practice of thinking through forms ultimately identifying ourselves in and with this very practice: this is the purity (Beauty), in itself mathematical, abstracted from the surface image in which Narcissus drowns. This is a type of order which spreads at the level of our still-forming brain on the basis of the autonomous development of a precise Neurogeometry, thereby approaching the definition of a Method, and it is in relation to the passing of the Method, and the ensuing new incarnation, that the renewed image in which Narcissus recognizes himself is defined. Orpheus dies through his own song, as it were, and the arising of the new Nymph is a consequence of this: hence the very turning of her eyes to the new image which emerges at the level of Nature's body as well as of her membrane. The final result of the song (like the innovative inner vision of fractals at the level of Klee's soul) consists in the fact that after the interrogation of the oracle, the artist's canvas or the notes of the divine singer come to posit themselves as the true response of the *experimentum crucis* opening up to the new incarnation and model. Actually, as we have already seen, the universe relating to the properties of numbers is far vaster than it can appear at the level of the image in which Narcissus self-encloses. A model characterized by recursive ordering, for example, can accommodate only a number of aspects connected with our ability to perform operations with numbers, even though it certainly illustrates some of the essential performances in which we recognize ourselves. Like Narcissus we close ourselves in a recursive frame, affirming our complete realization within it (Church's thesis), although with respect to specific operations as they have been calculated and verified, and we thus petrify with respect to the intended model which is realized semantically with reference to a universe in which numbers come to be directly identified as objects (subtending our intuitions-abilities). These numbers, while in themselves "reduced" (to pure notations), still allow us, within a particular level, to express a content with which we can identify while becoming possessed by truth. Here, however, we meet the imperious limit referred to by Skolem: at the level of the model we can, in effect, come to "see" as non-countable what is actually countable. Thus we come to be enclosed within an ineradicable "relativity". We could also, however, utilize non-standard models: it is our own choice to enclose ourselves, like Narcissus, in the intended model. When this happens, however, many of our abilities which are effective at the computational level will be removed, although we acknowledge, for instance, the usefulness of infinite numbers and make recourse to them. Closure within a model means surface closure: petrifaction consists in making the relative into an absolute able to allow us access to operations in themselves "safe". Self-enclosure requires reflexivity + Tennenbaum + finiteness, opening up, however, to infinitary instruments it extends the horizon and obliges us to take the road of abstraction (that of Marsyas): hence, for example, the need each time for an increasingly complex examination of unexplored order types. Within this framework relativity concerns the fact that the axioms of first-order set theory fail to capture univocally the actual mathematical practice in action and, therefore, the intended model at play. As we have seen, according to Skolem, PA only captures the structure of natural numbers to a relative extent, requiring recourse to more appropriate formalisms including, for example, as we have just said, infinitistic methods. Limiting ourselves to contemplating such a move, however, is one matter: preparing the new conception through sampling and petrifaction, thus opening up to a simulation work providing useful suggestions for possible renewed choices is somewhat different. In Henkin semantics the non-standard occurs when the quantifiers are chosen so as to vary above a finite set of domain subsets. According to Henkin, once provided with non-standard semantics SOL has the same properties as FOL. The intended-standard dialectics serves to extend the confines of petrifaction and provide suggestions regarding the Way of abstraction with a view to capturing new aspects of the mysterious life of numbers. The growth is ours, but it extends within the borders of the Other which comes into play at this level. The new Narcissus to be born will refer to a broader mathematical practice and properties deriving from the individuation of a renewed stage at the many-sorted level. Once again, Nature reveals itself as Language in action written in mathematical characters, albeit in relation to different orders and more sophisticated mathematical practices. A language, in particular, that concerns a canalization process relative to an intentional information that self-organizes with respect to three interconnected primary dimensions: capacity, control and sharing.

When new irruption occurs through the passing of the Method, and in connection with the ongoing flaying, the result will be not simple devastation, but necessarily, as in a Rubens' celebrated painting (and in a film by L. von Trier's), the presence of a quacking duck together with Mercury (P.P. Rubens, "Landscape with Philemon and Baucis", Vienna, Kunsthistorisches Museum and L. von Trier, "Nynphomaniac"). A new mathematical practice can then find its definition within a renewed model in relation to a new order type. The membrane will now be articulated with respect to a new level of the many-sorted case, and will be connected to a multidimensional composition of abstract machines linked in their turn to the realization of specific infinitary processes. At this level, the numbers (i.e. the system related to them) constitute one of the primary modalities through which the canalization process each time comes about in order to fix a possible conception. They, in perspective, pave the way for the inspiration which will now abound through reference to the simulation structures, ultimately allowing the intervention of a more general Method. There thus occurs the passing of this same Method at the level of the soul, as modulated on the basis of a renewed introduction of precise mathematical properties and relations (that is to say, in accordance with the notes of the tune piped by Pan). It is this passing which makes way for Apollo's new song by means of Orpheus, producing, too, the animating of new Nature in accordance with a renewed mathematical guise. A new language and new incarnation will emerge. A new membrane-nucleus correspondence will similarly be possible, with the identifying of a new intended model. The problem is how to enclose the infinite so as to make the right choices, not merely by producing simple multiplications but by cutting, forging with flame, and plumbing depths: actually only through this kind of selection and cutting can worlds and universes unfurl on our very flesh in a significant way. In accordance with a Wittgensteinian perspective, for example, as seen above, the standard comes to be enclosed within the confines of an in-depth "intentionality" allowing the hero to recognize and reflect himself in a model characterized by a recursive ordering. This closure, however, is the result of a series of limitation and reflection operations partly performed by means of the intervention of a targeted software. Meaning, in its turn, encloses itself in intentionality, the coincidence with the government in action reinforcing the confines of the model. The surfacing membrane can thus reveals itself in harmony with a nucleus which nests in its profundity. When this happens the Goddess self-nests completely within the stone book which opens as in a phantasmagoria or film (the mirror-book scrolled by the "theory" of Nature). She adorns herself for herself alone (albeit at the level of the Annunciation) and is burned by the Spirit at her nucleus but in accordance with a targeted recourse to the action of concepts: hence the birth of the Lord of Garlands as adeptness in plot elaboration and invention of possible worlds. The meaning which hitherto had been government is now use and opens up to real productivity, yet through its own death as a body and its "assumption" in the Heaven, as wondrously conceived by Titian in his mythological paintings. To the death of meaning (at the foot of the inverted cross) corresponds the death of the Father after the devastation and the abandon (as it emerges at the level of the same *experimentum crucis*). The God perishes and abandons, and light is withdrawn from meaning at the cross's foot. This is the destiny of Marsyas but also the abstract space in which the due recovery takes place. According to the unpredictability of the recovery path and to the audacious simulation activity deployed at the level of its preparation there can follow however the splitting and the correlated resurrection, with the assumption in the Heaven of meaning as Ariadne (and renewed software). Narcissus paves meaning with life but is in his turn closed within the truth. Unpredictable cuts will then be made and new orders revealed. The correct non-standard models must be recognized to give voice to true emergence, to the point of acknowledging the autonomous birth of new predicates and new mathematical practices.

References

1. Machover, M. (1996). *Set theory, logic and their limitations*. Cambridge.
2. Wang, H. (1970). Skolem and Gödel. *Nordic Journal of Philosophical Logic, 1*(2), 41.
3. Skolem, T. (1933). Uber die Unmöglichkeit einer volistandigen Charakterisierung der Zahlenreihe mittels eines endlichen Axiomensystems. *Norsk Matematisk Forening, Skrifter,* 73–82. Reprinted in Skolem, T. (1970). Selected Works in Logic. Oslo (pp. 345–354).
4. Skolem, T. (1941). Sur la portée du théorème de Löwenheim-Skolem. In *Les Entretiens de Zurich* (pp. 25–47), December 6–9, 1938. Reprinted in (Skolem 1970, pp. 455–482). [p. 475].
5. Gödel, K. (1934). Besprechung von "Uber die Unmöglichkeit einer vollstandigen Charakterisierung der Zahlenreihe mittels eines endlichen Axiomensystems. *Zentrallblatt fur Mathematik und ihre Grenzgebiete,* 2, 3. Reprinted in Gödel, K. (1986). *Collected Works*. New York. [p. 379].

6. Skolem, T. (1941). Sur la portée du théorème de Löwenheim-Skolem. In *Les Entretiens de Zurich* (pp. 25–47), December 6–9, 1938. Reprinted in (Skolem 1970, pp. 455–482). [p. 471].
7. Skolem, T. (1941). Sur la portée du théorème de Löwenheim-Skolem. In *Les Entretiens de Zurich* (pp. 25–47), December 6–9, 1938. Reprinted in (Skolem 1970, pp. 455–82). [p. 471–72].
8. Di Giorgio, N. (2010). *Non-Standard Models of Arithmetic: A philosophical and historical perspective* (M.Sc. thesis). University of Amsterdam. https://www.illc.uva.nl//MoL-2010-05.text.pdf.
9. Wang, H. (1996). Skolem and Gödel. *Nordic Journal of Philosophical Logic, 1*(2), 119–132. [p. 122].
10. Wang, H. (1996). Skolem and Gödel. Nordic Journal of Philosophical Logic, 1(2), 119–132. [p. 122–3].
11. Machover, M. (1996). Set Theory, logic and their limitations. Cambridge. [p. 282].
12. Gitman, V. (2015). An introduction to nonstandard models of arithmetic. In *Analysis, logic and physics seminar*, April 24. Virginia Commonwealth University. [p. 2].
13. Carnielli, W., & Se Araujo, A. (2012). Non-standard numbers: A semantic obstacle for modelling arithmetical reasoning. *Logic Journal of the IGPL, 20*(2), 477–485.
14. Hamkins, J. D. (2018). *Nonstandard models of arithmetic arise in the complex numbers.* http://jdh.hamkind.org/nonstandard-models-of-arihtmetic-arise-in-complex-numbers/.
15. Baez, J. (2018). *Nonstandard integers as complex numbers.* Azimuth. https://johncarlosbaez.wordpress.com/.../nonstandard-integers-as-complex-numbers/. [p. 1].
16. Baez, J. (2018). "Nonstandard Integers as Complex Numbers", Azimuth, [https://johncarlosbaez.wordpress.com/.../nonstandard-integers-as-complex-numbers/], [p. 1–2].
17. Paris, P., & Harrington, L. (1977). A mathematical incompleteness in Peano arithmetic. In J. Barwise (Ed.), *Handbook for mathematical logic.* Amsterdam, Netherlands: North-Holland.
18. Benacerraf, P. (1965). What numbers could not be. *The Philosophical Review, 74,* 47–73. [p. 50].
19. Halbach, V., & Horsten, L. (2005). Computational structuralism. *Philosophia Mathematica, 13*(III). [p. 178].
20. O'Regan, J. K., Myin, E., & Noe, A. (2004). Towards an analytic phenomenology: The concepts of "Bodiliness" and "Grabbiness". In A. Carsetti (Ed.), *Seeing, thinking and knowing.* Dordrecht. [p. 104].
21. Quinon, P., & Zdanowski, K. (2006). *The intended model of arithmetic. An argument from Tennenbaum's theorem.* http://www.impan.pl/_kz/_les/PQKZTenn.pdf.
22. Lawvere, W. (1969). Diagonal arguments and cartesian closed categories. *Lecture Notes in Mathematics, 92,* 134–145.
23. Dean, W. (2013). Models and computability. *Philosophia Mathematica (III), 22*(2). [p. 63].
24. Dean, W. (2013). Models and Computability. Philosophia Mathematica 22(2): 143–166. [p. 63].
25. Putnam, H. (1980). Models and reality. *The Journal of Symbolic Logic, 45,* 464–482. [p. 479].
26. Quinon, P., & Zdanowski, K. (2006). *The intended model of arithmetic. An argument from Tennenbaum's theorem.* http://www.impan.pl/_kz/_les/PQKZTenn.pdf. [p. 2].
27. Carsetti, A. (1990). Algorithmes, complexité et modèles. *La Nuova Critica*, 15–16: 71–100.

Chapter 4
Regulatory Logic, Algorithmic Information and General Semantics

Abstract As H. Atlan and A. Carsetti remark in a natural self organizing system (a biological one, characterized by the existence of cognitive activities) the goal has not been set from the outside. What is self-organizing is the function itself with its meaning. The origin of meaning in the organization of the system is an emergent property. In this sense the origin of meaning is closely connected to precise linguistic and logical operations and to well defined procedures of observation and self-observation. These operations and procedures induce, normally, continuous processes of inner reorganization at the level of the system. The behaviour of the net, in other words, possesses a meaning not only to the extent that it will result autonomous and to the extent that inside it we can inspect a continuous revelation of hidden properties, but also to the extent that it will result capable of observation and self-observation as well as intentionally linked to a continuous production of possible new interpretation acts and reorganization schemes. The state space relative to these observational, self-observational and intentional functional activities cannot be confined, in this sense, on logical grounds, only within the boundaries of a nested articulation of propositional Boolean frames. On the contrary, we will be obliged, in order to give an explanation, within a general theory of cellular nets, for such complex phenomena, to resort to particular informational tools that reveal themselves as essentially linked to the articulation of predicative and higher order level languages, to the outlining of a multidimensional information theory and to the definition of an adequate intensional and second-order semantics.

4.1 Regulatory Logic and Self-organization in Biological Complex Systems

With respect to the Theoretical Biology and the newborn Metabiology, scientists often take up, from an epistemological point of view, a Kantian position as regards both the nature and limits of scientific knowledge and the relation between knowledge and reality. In accordance, for instance, with E. R. Dougherty and M. L. Bittner we have, first of all, to recognize that there are things-in-themselves outside the direct structure of our scientific knowledge, which is constrained by the limits of

A. Carsetti, *Metabiology*, Studies in Applied Philosophy, Epistemology and Rational Ethics 50, https://doi.org/10.1007/978-3-030-32718-7_4

sensibility. "The observations are tied to scientific knowledge, which is constituted by mathematics, and the validity of the knowledge is relative to predictions based upon the mathematics and tested via operational definitions" [1].

Biology, in general, concerns living organisms. It depends upon Physics and Chemistry. But the subject matter of Biology concerns the operations performed by the cell in its pursuit of life, not the molecular infrastructure that forms the physiochemical *substratum* of life. "The activity within a cell is much like that within a factory. In the latter, machines manufacture products, energy are consumed, information is stored, information is processed, decisions are made, and signals are sent to maintain proper factory organization and operation" [1].

According to their opinion, the hardware units within a factory, whether mechanical, electrical, or chemical, do not constitute the factory. These require specialized knowledge to build and are necessary for the factory to function but, in and of themselves, they simply compose a collection. They become part of a factory when their functioning is organized and regulated according to programs that integrate their activities in such a way as to produce the desired products and maintain their proper functioning within the overall operation of the factory. What constitutes the factory as an entity is the regulatory logic (the set of logical programs) that controls the dynamics of the factory. The same can be said for a cell if we strip away that which is purely physical, if, that is to say, we make exclusive reference only to the "living" software in action in accordance, for instance, with Chaitin's last suggestions. "Cells have—they write—a similar approach to managing their responses to critical changes that could lead to death or substantial damage to the organism. These actions fall in the category of stress responses, system-wide alterations that deal with various environmental insults and cellular malfunctions. In these situations, many processes may need to be halted and many others instituted. … In regulatory processes, where chains of signals are used to induce systemic changes in the functions a cell is currently performing, the presence or absence of particular gene products that mediate the turning on and off of the production or function of the gene products targeted by the regulatory action can be used to specify whether one or another particular bank of genes will be acted on and whether the action will be an induction or cessation of their action" [2].

From a general point of view, if the regulatory logic can handle an asynchronous system, there is significant gain in efficiency; moreover we well know that a cell (as well as a factory) requires redundancy to keep operations running smoothly. However, as von Forester, Pugachev, Atlan, Carsetti have remarked, in order to assure true redundancy we have to refer to systems capable of incorporating elements for automatically adjusting particular parameters according to an analysis of input and output data. Complex systems of this kind appear also capable of adapting themselves completely at each instant to the results of their analysis of external conditions. These systems are said to be *self-organizing*. Self-organization is more than simple redundancy. It allows a system to reconfigure itself to achieve optimal (practically, close to optimal) performance under varying conditions. This reconfiguration appears as linked, at the biological level, to the opening up of the living system and the sudden development of specific virtualities.

We well know that the cells use redundancy and parallelism to deal optimally with damage and malfunction. It is also evident, however, that in cells, not only are there multiple inputs involved in a decision, there are alterations in the hardware components that interpret the inputs, making the responses context-sensitive. This occurs, for example, when a controlling gene is capable of acting to produce a certain set of regulatory results only when its actions are interpreted by a particular set of gene products that are variably present. According to Dougherty and Bittner: "This is a widespread problem in gene control, since in the cell, every gene being expressed is regulated by other genes and frequently there are multiple regulatory conditions for a gene to be expressed, so that any set of samples is likely to have many genes being controlled by different genes in different samples, thereby making simple correlation a poor way to identify regulatory connections … there are two basic operational issues concerning a factory: characterization and control of its operations. First, we want to characterize the input and output of the factory; second, we want to organize the operations so as to achieve optimal (or at least satisfactory) performance. Both characterization and control require a suitable conceptualization of the factory. Such a conceptualization must be mathematical for two reasons. First, characterization and control involve relations among the components and mathematics provides a relational language, and second, mathematics provides a language in which complexity can be represented in such a way as to be amenable to human analysis. Moreover, not only are complex systems beyond ordinary intelligibility and intuition—indeed, their performance is often highly counter-intuitive—but they typically cannot be fully represented mathematically because there are too many relations and, even should one achieve a very precise and highly involved mathematical description, it may well be intractable relative to solutions of the problems of interest, such as optimizing some set of relation within the system" [3].

In their opinion, for both factory and cell, the regulatory logic determines the relations between the physical structures within the system and between the system and its environment. "By regulatory logic we do not simply refer to simple binary deterministic logic but to mathematical functions, perhaps binary in nature, that provide operational control within the framework of random processes. The roles of regulatory logic in the factory (or complex machine) and the cell are congruent because the key to the characterization of this logic lies in communication (between components) and control (of components)—that is, in systems theory, which therefore determines the epistemology of the cell" [4]. In this sense, randomness, on the one hand, and logical programs, on the other hand, appear to be at the hearth of both life and cognition.

To illustrate these epistemological points, Dougherty and Bittner consider a regulatory model that incorporates latency, context-dependence, distributed regulation, multivariate gene interaction, and stochasticity. Because regulation is parallel and distributed, if one views the cascade of activities resulting from the action of a single regulatory gene, both the strength and specificity of subsequent activities in the cascade may be expected to diffuse through subsequent steps in the cascade. As the regulatory effects propagate, they are progressively modified or limited by interactions with other factors modulating gene transcription.

"Biology studies relations between molecules (chemical structures), not the molecules or the forces between molecules. The recognition that biological knowledge concerns regulatory logic and the consequent intra-cell operational organization of molecular structures, as well as, by extension, inter-cell organization, entails the concomitant recognition that biological systems, in their extraordinary complexity, are beyond everyday intelligibility and intuition. Moreover, it facilitates answers to the two fundamental epistemological questions raised in the Introduction: (1) What form does biological knowledge take? (2) How is biological knowledge validated? The question as to form relates to the type of mathematics involved in modeling the relations that characterize regulatory knowledge. This depends on the nature of the relations being considered; however, the general mathematical framework will be formed within the theory of stochastic multivariate dynamical processes. Validation depends on the mathematical model characterizing the knowledge and, since this knowledge concerns operational regulation, validation will involve operational predictions derived from the mathematical model" [5]. Let us underline, however, that these predictions involve, at the biological level, the "intelligent" choices performed by an epistemic agent. The dialectical pairing between creativity and meaning is also linked to the incessant identification of specific cognitive tools, to the capability on behalf of the epistemic agent to individuate the correct "languages" in order to model mathematical functions able to account for the biological phenomena of interest.

As we have just said, in their paper Dougherty and Bittner assume, from an epistemological point of view, a Kantian position: moreover, they primarily refer to simple systems (considered as elements of a collection) and complex arrangements of these systems in order to outline specific forms of self-organization at the level of these very systems. But, what happens if we assume as "our" starting point complex coupled systems instead of simple systems? As we said in the first chapter, at the level of natural evolution, when observing phenomena which are neither purely ordered nor purely random, but characterized by a plot of sophisticated organization modules, we are normally observing an intermediate situation between a complete absence of constraints on the one hand, and the highest degree of redundancy on the other. In this sense, optimal organization should be considered as an effective compromise (Schrödinger) between the greatest possible variability and the highest possible specificity. Given the deep structure underlying the surface message, this compromise inevitably articulates itself in accordance with a dynamical dimension proper to a self-organizing process, i.e. a process on the basis of which changes in organizational modules are apparently brought about by a program which is not predetermined, but produced by the encounter between the realization of specific virtualities unfolding within the evolving system, and the action, partly determined by specific coagulum functions, of generative (and selective) information fluxes emerging at the level of the reality "external" to the system. In principle, this process may for a variety of factors cause a progressive reduction in the conditions of possible high redundancy proper to the initial phase, together with a successive sudden increase in symbolic potential variability. This will, therefore, allow a further extension of the internal regulatory factors' range of action, in conjunction with the appearance of new constraints and regenerated organizational forms, so that the increase in variability will be matched by increased specificity and reduced disorder, and with no paradox or contradiction involved. It is precisely this near-contextual increase in variability and specificity that we refer to in speaking of an internal "opening up" of a coupled system furnished with self-organizing modules. In other words, when at the level of life specific thresholds are crossed, the bases of variability are extended, while, simultaneously, the conditions favorable for extremely complex organizational phenomena begin to be realized.

The variability-specificity dialectics characterizing self-organization (at the level of living systems) appears, on a more general level, to be strictly connected to the evolution of an amply- acknowledged function: thermodynamic entropy. We know, for example, how, under certain conditions relative to the articulation at the co-evolutionary level of a specific self-organization process, physical entropy can increase as disorder simultaneously decreases, particularly when the number of microstates is not constant and increases more rapidly than the number relative to S (entropy). According to Carsetti [6], what this evinces, however, is the essential relation, at the entropic level, between the cognitive agent's information about the system, on the one hand, and the increase in entropy, on the other hand. Thus, to the coupled evolution of the system and the environment, and the variability-specificity dialectics, a further element must be added, the autonomous (and cognitive) agent considered as a factor of primary reference, the functional role of which it is necessary to clarify. In terms of effectiveness, the calculations of entropy would seem, within the realm of statistical mechanics, to depend on what is known of the system and what measurements can be performed on it. The probability distribution changes very essentially if the distribution is to be applied to a different number of accessible states from that originally calculated. In this case the information can increase entropy in real terms. Let us consider, as an example, the common definition of entropy as a measure of disorder and let us point out, in accordance with Bais and Doyne Farmer [7] that the thermodynamic entropy is indifferent to whether motions are microscopic or macroscopic: it only counts the number of accessible states and their probabilities. Thus, if we define probabilities that depend only on the positions of the particles and not on their velocities we obtain entropy associated with a new set of probabilities which we might call "spatial order entropy", an entropy that articulates quite differently from the thermodynamic entropy.

Given that, according to the theoreticians of "coarse graining", a cognitive agent is unable to determine the details of microscopic dynamics beyond a certain scale of size, to describe events on a smaller scale it will be necessary to perform the necessary calculations on the basis of average behavior. It is the average of total probability for groups of microscopic states which is usually calculated, and this average probability is attributed to each microscopic state, maintaining constant $n(t)$, the number of accessible states. It is to this type of manipulation, the nucleus of coarse graining, that entropy increase in statistical mechanics is to be attributed according to Landsberg, for example. Here, in his opinion, we can identify the real emergence of what is virtually time's arrow.

Thus, when we make reference to coupled systems and to their possible "opening up" (to the effective realization of the afore mentioned "compromise"), we have to take into account a particular series of functional dichotomies: depth versus surface, variability versus specificity, unfolding versus coagulum, microscopic versus macroscopic, invariance versus morphogenesis, denotational versus operational, form versus information and so on. This multiple and twisted dialectics finds its nucleus in the operational and indissoluble link between life and cognition. Cognition is ubiquitous across the living state, it is necessary for life, even if it is not sufficient. Actually, an organism represents a highly patterned outcome of path-dependent variation, selection, and interaction, in the context of an adequate flow of free energy (information) and an environment fit for development. Complex cognitive processes operating at the level of a living system record, instantiate, guide and nourish the evolutionary and developmental trajectories characterizing this very system.

In this sense, the ubiquity of cognitive process at all levels of scale and organization within any living organism leaves open to evolutionary exaptation the noise inherent to information transmission. A principal outcome of such exaptation will be the construction of some punctuated global transmission mechanisms at various scales that instantiate the "arches" proper to the arising new forms of life. If we accept the thesis supported by Carsetti according to which gene expression should be partially inserted in the realm of complex linguistic behavior for which context imposes meaning [8], we can immediately realize, as R. Wallace maintains, that this analogy between gene expression and language production is useful both as a fruitful research paradigm and also, "given the relative lack of success of natural language processing by computer, as a cautionary tale for molecular biology" [9]. Hence the essential role played by operational meaning at the level of living systems. The computational processes that inhabit the living factory must be related, in an essential way, to their contexts and their meanings. However, given that we do not possess an adequate outlining concerning the semantics of biological (and natural) languages, our first aim should concern the invention and the development of new mathematical tools in order to build this kind of semantics. As we shall see, this development, in turn, is part of the very process concerning our self-organization as living and cognitive systems.

4.2 General Semantics, Algorithmic Information and Second-Order Logic

As we pointed out in the second chapter, at the biological (and creative) level, the original, developmental and selective source, while transmits and applies its message, constructs its own structure. The transmission content is represented by the progressive "revelation" by forms of the very source, of the self-organizing (and emotional) "instructions" concerning, each time, its actual realization at the surface level and its primary operational closure. With respect to this biological landscape, meaning assumes precise contours insofar as it is able to express itself through the ruler, but in accordance with the action of the *telos*, thus helping in realizing a precise "inscription" process within the space belonging to the Temple of life. As we have said, the Temple's columns flank the passage to the standard (in a precise mathematical sense, in accordance for instance with the intuitions of Reeb's school [10]), marking the successive adjunction of different observing systems who posit themselves as the real supporters of the ongoing construction. These acts of observation, then, are not linked only to decision and conventions, but more centrally to the unfolding of a process of incarnation and to the dialectical relationship between ruler and coder as well as to the continuous emergence, at the evolutionary level, of new formats of sensibility.

It is through the unfolding of this continuous metamorphosis process that a coupled system may come to recognize (and pursue) the secret paths of the intentional information characterizing its intrinsic development as programmed by natural selection. The final result is a source that assumes a replicative capacity commensurate with a precise invariance and the progressive constitution of specific forms inhabiting life. This particular process develops in a Darwinian landscape in accordance with the principles which govern natural evolution. With respect, however, to these principles, we feel, today, the need to better understand how and in what terms the processes of self-organization and the dialectic pairing between creativity and meaning determine, in an effective way, the course of evolution. We aim, in other words, to grasp mathematically the *Sinn* that governs, on a deep level, natural evolution. As we have seen in the second chapter, in some recent important papers and books, G. Chaitin has revisited the scientific status of evolution theory, as stated by Darwin, by asking the following questions: is it possible to give a mathematical proof of natural evolution? is it possible to make reference not only to a logic that governs natural reality from within, but to program libraries able to express forms of self-organization with an evolutionary character? is it possible to give an explanation concerning the ultimate reasons that lead the living systems to evolve? As we already said before, in order to give a first answer to these difficult questions Chaitin took his first moves from a revisitation of his earlier works on Algorithmic Information Theory (AIT), the theory, namely, to which the great scholar has devoted his entire life.

From a general point of view, traditional information theory states that messages from an information source that is not completely random can be compressed. Starting from this statement, Chaitin in the seventies outlined his general idea of randomness and introduced that conceptual notion of complexity that is at the core of AIT: if lack of randomness in a message allows it to be coded into a shorter sequence, then the random messages must be those that cannot be coded into shorter messages. Compared to this conceptual framework, we have also seen in the previous chapters how a very efficient systematic choice as regard the possible simulation of the fitness function, can be represented by Chaitin's Ω number. Chaitin uses Ω_k to define an organism $\mathbf{O}_k$ and a mutation M_k at time-step k, as well as the fitness function W. Namely, an organism is defined by means of the first $N(k)$ binary digits ω_i of Ω_k. The mutation acts on the organism by trying to improve the lower bounds on Ω. These mutations represent challenging an organism to find a better lower bound of

Ω which amounts to an ever increasing source of knowledge. In this way, according to Chaitin's model, evolution and randomness appear strictly intertwined.

Let us note, however, that the statement that chaos is deterministic randomness is even more precisely formulated if the laws of chaos are predicated on the basis of the traditional undecidability theory. For example, it is possible to demonstrate how in specific dynamical systems, chaos is simply the effect of the undecidability proper to a recursive, system-related algorithm. At an even more general level, we can state that deterministic, chaos-generating systems may be considered as set-ups able to act as models (in a logical-mathematical sense) for formal systems (algorithms) within which it is possible to trace the trajectory of a mathematically realized randomness. In other words, it is possible to consider the randomness defined at an algorithmic-informational level, as the counterpart and theoretical foundation of the physical randomness connected to the articulation of well-defined processes—i.e. set-ups. This last kind of randomness, however, must be considered as absolute, not only because it satisfies the requisites of objectivity-coherence and intrinsic incompressibility, but also because the cognitive agent is able to define it principally by stating the limitation theorems concerning the formal system within which randomness is recognized as such, particularly with regard to that specific system which, at an algorithmic and programming level, assists, as we have just said, in the "guided" construction of the corresponding set-up. Indeed, when the cognitive agent becomes aware of the fact that a particular axiomatic system allows him to define a given number as random, but is still unable to resort to his formal tools in order to calculate that number or to permit a proof of its purely casual nature, s/he sees a concrete foreshadowing of an absolutely random string which, out of all possible messages, will be that which is most redolent of information precisely because all redundancy has been eliminated. It is exactly in this sense that Ω can be directly considered as the Pillars of Hercules of Reflexivity [11].

Chaitin is perfectly right to bring the phenomenon of evolution in his natural place which is a place characterized in a mathematical sense: Nature "speaks" by means of mathematical forms. Life is born from a compromise between incompressibility and meaning, on the one hand, and, on the other hand, is carried out along the ridges of a specific canalization process that develops in accordance with computational schemes. It is with respect to this particular channeling that the fitness must be calculated according to the deployment of precise functions. Hence the emergence of those particular forms (linked to specific development languages) that are instantiated, for example, by the Fibonacci numbers, by the fractal-like structures etc. that are ubiquitous in Nature. As observers we see these forms but they are at the same time inside us, they pave of themselves our very organs of cognition.

In the light of these considerations, let us remember once again that the pure (classical) reflexive models which are not open to a creative exploration articulating at the second order level are not able to account for true creativity and real metamorphosis because they do not take into consideration the dialectical pairing of creativity and meaning as well as the emergence processes living at the level of meaning. They do not loosen the knot of the intricate relationships between invariance and morphogenesis and do not arise in relation to the actual realization of a specific embodiment. So to speak, they remain close in their skeleton. Hence the importance of making reference to theoretical tools more complex and variegated as, for instance, non-standard mathematics and complexity theory, in order to provide an adequate basis for the afore mentioned extension. Let us resort to an exemplification: the von Koch curve is an eigenform, but it is also a fractal. However, it can also be designed utilizing the sophisticated mechanisms of non-standard analysis. In this last case, we have the possibility to enter a universe of replication, which also opens to the reasons of real emergence. At this level, the growth of the linguistic domain, the correlated introduction of ever-new "individuals" (as well as of new mathematical entities) appears strictly linked to the opening up of meaning and to a continuous unfolding of specific emergence processes with respect to this very opening. Hence the need for the introduction of precise evolutionary parameters, the very necessity, in general, to bring back the inner articulation of the eigenforms not only to the structures of "simple" perception but also to those of intentionality. In order to capture the meaning of this last statement, let us to add some more considerations with respect to the inner articulation of the eigenforms. As we have just said, the von Koch curve is an eigenform: as such it is a limit of a process. But we may wonder: does this limit exist? is there an invariance property for the Koch curve? Actually, now we have a definite answer with respect to these important questions [12]. But what happens if we decide, in analyzing our curve, to explore another formalism able to introduce new "elements", to deal, for instance, with infinitesimal and infinite numbers thus introducing a set "larger" than **R**? We are obviously obliged to enter the realm of non-standard analysis. At this level the set **R** of standard numbers is a subset of ***R**, the set of hyper-real numbers. ***R** contains infinitely small and infinitely large numbers. ***N** will denote the set of hyper-natural numbers. Let us generalize the usual fractal by introducing a curve F_ω in ***R** parameterized by an hyper-real number $x^* \in {}^*[0, 1]$: $x^* = x_1/p + \ldots + x\,p^n + \ldots + x_{\omega-m}/p^{\omega-m} + \ldots + x_\omega/p^\omega$, where $\omega \in {}^*\mathbf{N}_\omega$. F_ω is not a fractal in the non-standard sense (since the fragmentation is *limited up to ω) but its standard part is identical to the usual fractal: $F \equiv st\,(F_\omega)$. By using an infinitely great magnifying power, F_ω can be drawn in an exact way: the fractal is no more a limit concept. Making reference to this particular and very simple "landscape", we can easily realize that the constraints imposed by specific selective pressures (operating in ambient meaning and articulating in accordance with suitable non-standard procedures) at the level of the dynamics of an original cellular (dissipative) automaton can, actually, permit a more complex canali-zation of the informational fluxes at stake. In particular, they can allow the unfolding of silent potentialities, the full expression of generative principles never before revealed and, consequently, the effective expression of new autonomous processes of production of varied complexity.

When we are in second-order logic, but we make essential reference to non-standard interpretations and allow structures with non-full relational universes, quantification only applies for the sets and relations that are present in the structure. In the general structures of L. Henkin (1950), for instance, we put in the universes all sets and relations that are parametrically definable in the structure by second-order formulas. In this sense, it is not surprising that the set of standard numbers is not definable by a second-order formula in a structure having non-standard numbers.

What is important to stress again is the fact that hidden in the structure some specific relations exist, some "rules" (second-order relations) that cannot be defined as relations among individuals, but are utilized to define first-order relations (i.e., relations among individuals). As a result, we obtain a particular structure where the n-ary relation universe is a proper subset of the power set of the n-ary Cartesian product of the universe of individuals. So, whereas in the standard structures the notion of subset is fixed and an n-ary relation variable refers to any n-ary relation on the universe of individuals, in the non-standard structures, on the contrary, the notion of subset is explicitly given with respect to each model. Thus, in the case of general structures the concept of subset appears directly related to the definition of a particular kind of constructible universe, a universe that we can explore utilizing, for instance, the suggestions offered by Skolem (cf. his attempt to introduce the notion of propositional function axiomatically (1960)) or by Gödel (cf. Gödel's notion of constructible universe (1972)).

From a more general and philosophical point of view, we can say that at the level of general structures, the relations among individuals appear as submitted to a bunch of constraints, specifications and rules having a relational character, a bunch that is relative to the model which we refer to and that acts "from the outside" on the successive configurations of the first-order relations. In other words, in the universes of any second-order frame Ψ there are only relations among individuals, but it is no longer true that all the n-ary first-order relations on Ψ are into Ψ.

These hidden relations, these particular "constraints", play a central role with respect to the genesis of our models. In particular, let us remark that as a consequence of the action performed by these constraints, the function played by the individuals living in the original universe becomes more and more complex. We are no longer faced with a form of unidimensional relational growth starting from a given set of individuals and successively exploring all the possible relations among individuals, according to a pre-established surface unfolding of the relational texture. Actually, besides this kind of unidimensional growth, further growth dimensions reveal themselves at the second-order level; specific types of development that spring from the successive articulation of the original growth in accordance with a well defined dialectics. Such a dialectics precisely concerns the interplay existing between the first-order characterization of the universe of individuals and the whole field of relations and constraints acting on this universe at the second-order level. As a result of the action of the rules lying at the second-order level, new dimensions of growth, new dynamic relational textures appear. Contemporarily the original universe of individuals changes, new elements grow up and the role and nature of the ancient elements undergo a radical transformation. In this sense, the identification of new growth dimensions necessarily articulates through the successive construction of new *substrata*. The aforesaid dialectics reveals itself as linked to the utilization of specific conceptual tools: limitation procedures, identification of fixed points, processes of self-reflection and self-representation, invention of new frames by "fusion" of previously established structures, coagulum functions etc.

What we have remarked until now permits us to understand more deeply the ultimate sense of Henkin's conceptual revolution. As M. Manzano (1996) correctly remarks, Henkin arrived to prove the completeness theorem for type theory, "... by changing the semantics and hence the logic. Roughly presented, the idea is very simple: The set of validities is so wide because our class of standard structures is too small. We have been very restrictive when requiring the relational universes of any model to contain all possible relations (where "possible" means in the background set theory used as metalanguage) and we have paid a high price for it. If we also allow non-standard structures, and if we now interpret validity as being true in all general models, redefining all the semantic notions referring to this larger class of general structures, completeness (in both weak and strong senses), Löwenheim-Skolem, and all these theorems can be proved as in first-order logic" [13].

In this sense, in accordance for instance with Németi's [14] opinion, standard semantics is not logically adequate because it does not include all logically possible worlds as models. On the contrary, in Henkin's general semantics [15] many "hidden" possibilities are progressively taken into consideration as possible models. We can have, for instance, models with or without GCH (generalized continuum hypothesis). Things are really different in the case of standard semantics.

The conceptual importance of the discovery of non-standard models can be well understood if we try to elucidate a precise dialectical aspect characterizing the link between non-standard models and undecidable propositions. Actually, even if Gödel's theorems indicate to us how to build certain formulas which are shown to be true but unprovable, however, as Henkin remarks, there is no general method indicated for establishing that a given theorem cannot be proved from given axioms. Such a method is exactly supplied by the procedures of constructing step by step non-standard models for number theory in which "set" and "function" are reinterpreted. These procedures show us that in order to model thought processes in an adequate way we must explore the non-standard realm on the basis, first of all, of the identification of precise fixed points and the tentative definition of new kinds of universes. For example, we know that, in accordance with Mostowski's results (1965), Gödel's famous undecidable proposition can be simply considered as a proposition that characterizes the class of natural numbers. If we refer this proposition to a system of non-standard numbers, it will no longer be valid. In this way, we can realize that, along our exploration, we are really driving specific "conceptual" stakes into the ground of an unknown territory and that this exploration articulates in a co-evolutionary landscape. At the level of this particular landscape constructing and discovering appear as dialectically interrelated.

In this way, we have the possibility to preserve the deep insights outlined by Chaitin (2013) relative to the mathematical *substratum* underlying biological evolution. At the same time, we also have the possibility to refer them to the realm of that dialectics between creativity and meaning that allows us not to close ourselves into the enclave of the first-order structures but to range in a much broader realm of functions also featured in accordance with the tools offered by non-standard mathematics. Thus, it will then be possible to take into account some general themes concerning, for instance, the role played by: (1) meaning in action; (2) the computational membranes at the level of the development of living beings; (3) those specific processes that determine that particular emergence of ever-new biological structures that distinguishes natural evolution; (4) the *telos* at the level of biological evolution. In a nutshell, we shall have the opportunity to enter the mystery concerning, at the biological level, the stratified and hierarchical development of the differentiation processes.

As stated above, the intentionality belonging to meaning manages, through the brain's channeling, to rise to the surface of the disemboding coder, constituting itself as membrane able to mirror in itself the coder's nesting by programs. Here is Marsyas as a Painter in action, a Painter who builds and tells-describes himself up to realize extroversion, thus offering the flank to the selection by the God who, through him, has had the opportunity to realize the recovery. Hence the irruption, but from here also the surfacing of a countenance understood as the emergent face of the Muse (as a result of the ongoing self-organization of meaning together with its creativity), a countenance that the Painter comes to experience following his own death by torture. Here is a sudden gash in the sky that opens to new thought (creativity). When, however, it is creativity that self-organizes together with its meaning, there is the emergence in the night of meaning of a hero's face: the face of Endymion. No longer a gash in the direction of a soul but a leaven in the direction of a body and a cyclical return. Hence the dance of the heroes that fills the night of the Goddess, a dance of which Endymion constitutes, precisely, the emblem. In place of the opening-tear relative to a soul (of the Work), we will now have the closure in a cyclical melody (dance) related to a body (of Nature). When the Muse breaks in, there is a gash relating to the emotion of the God and there is a creativity that presents itself, precisely, as enthusiasm in action. When the heroes rise and nest in a meaning that is channeled as an eternal and cyclical rule, we have instead the addition to the Temple by the bodies of the heroes in the context of their closure in a dance each time renewed under the guidance of the Goddess. Hence the possible conception: irruption by creativity and conception by meaning.

Starting from the irruption we will have the detachment of the Minotaur, starting from conception we will have the lurking of the Lord of the wreaths: the Goddess who received the offer of the cipher and who can, therefore, reflect herself with reference to the scanning carried out at the level of the dance of the heroes can, actually, come to conceive. Compared to both the new conception and the nesting process as performed by meaning, the final result is an extended arborization (Daphne) revealing itself at the surface level (in that life traverses it) as a design-web of connections unified by a living intentionality finally expressing itself as true harmony. This harmony is that of signs created by a "hand" which ramifies to give rise, through the carving operated (as instantiated, for example, by Rodin in a famous sculpture) to the impetuous irruption of the God and his emotion as this last comes to take shape through the action of painting and, in general, of the praxis of art. Here is a hand (the hand of Marsyas) that individuates (and perceives) itself, through the operated design, as the combined and productive features of meaning in action. Through the realized arborization meaning reaches the surface-life of pure abstraction and freezes in all its conceptual "elegance": membrane-mirror which accommodates and enfolds all things within itself including its own impenetrability.

At the level of the irruption, we are faced with a Painter who, as happens in Vermeer, can finally enter the scene in the framework of the "Action of Painting", thus leaving the absence and presenting himself in all the richness of his attributes. In order to find, however, new salvation by taking on a renewed purity, it will be necessary to gain in depth new incompressibility. The surface will have to be torn apart and the harmony of a new intentionality will have to come to be recovered: only in this way can a new soul come to reveal itself in the sky. Hence the necessary sacrifice on the part of the Painter as it is represented, for example, in Titian and Vermeer. The coder lurks for programs (software) along the abstraction giving rise to a new irruption, the ruler emerges as hardware along the incarnation giving rise to new conception. The coder leads to the manifestation of Clio as impetus while the ruler that emerges as hardware leads to both the pulse of Endymion and the new conception through the offer of the cypher.

Endymion, in fact, constitutes—represents in himself the pulse of the heroes along the course of the dance guided by Flora, a dance that presides over the cyclical return to life and that can come to be realized only on the basis of a successful adjunction. If Clio appears linked to the irruption as effected by the God, Endymion is linked to the act of conceiving by the Goddess: it is through the hero that particular scanning that allows the Goddess to conceive can come to be configured. Creativity selects and the Muse bursts: the identification of the score gives rise to the possible resonance of the harmony of a soul. Here is Vermeer which is selected and shaken: the painter having allowed the recovery can come to make experience of the impetus of the Muse. Marsyas-Painter, in effect, assures on his own skin the renewal of the God's creativity and the imperious manifestation of the face of the Muse, of the renewed emotion that comes, now, to turn his eyes. Narcissus, for its part, working for the offer of the cipher ensures both the renewal of conception and the silent manifestation of the rule according to an eternal return. Actually, the hero brings to fulfillment the incarnation to the extreme limit of petrifaction. The bursting of the Muse as a result of the self-organization enacted by meaning, will therefore correspond to a creativity that springs from the fulfilment of the Work and gives rise to the new surfacing of the God. Here is an action of painting which leads to the fulfilment of a life and which marks the elevation of a soul. The Painter (Vermeer) enters the Work from which he was always absent and comes out of life: his creativity now lives on the level of God. At the nesting of the Goddess following the self-organization fielded by meaning will come to correspond a creativity that self-organizes together with its meaning up to preside over the articulation of a new Nature. Hence the hero who pierces himself with reference to his own image. The observer returns cyclically to life thus entering the dimension proper to a dream (concerning a reality of stone: see Poussin and Proust) and placing himself as a witness of a particular type of existence understood as a cyclical return to the intact beauty of a body eternally enclosed in its identity. The meaning that concerns him will come, now, to pulsate in the eternal light of a Goddess scanned through a constellation of fixed points, a Goddess that turns out to be able to conceive the new Lord of the garlands: here, for instance, is the message forwarded by M. Antonioni through Zabriskie Point. Eigenvalues that support the outcrop and eigenforms that mark the nesting. Creativity emerges to itself, meaning is involved in itself. Anchorage to the ever-changing face of the Muse and anchorage to the ever-equal body of Endymion. Truth that breaks every time different in the light of creativity and life that repeats itself identical in the *Sylva* of meaning. Truth of the soul and life of the body. Truth and irruption, life and conception.

With respect to this framework, the hand that performs the simulation of the creative process up to the offer of the stool (as shown by Balthus in the painting: "*Grande composition au corbeau*") and which will then be selected by the Method, will give rise to an effective recovery with consequent birth of new creativity. Hence a life in the truth that will present itself as the *Via* and as a pure function, a function that can never, however, turn back on itself closing in its ivory tower if one wants to avoid the self-abandonment by the God. Hence the need for the God to continue to add ever new creators to himself. It is the artificial hand of the craftsman who comes to reveal itself as natural in the burning bush that underlies and witnesses the recovery in action by the God who can thus come to unfold his own intentionality in accordance with a renewed creativity. But for this to happen, meaning must come to self-organize along with its creativity in order to conceive. The artist manages to capture life in the truth by means of his Work to the extent that this same Work allows the God to express himself in the bush of creation and to leave the abandonment (of himself) following successful recovery (the recovery of which the Saints in Beato Angelico stand as witnesses at the level of their creative vision). If Marsyas supports the recovery by offering the correct stool thus coming to ensure the rising of the Nymph, as well as the surfacing in a renewed way by the God as *Natura naturata*, here is the actual possibility for the Minotaur-Lazarus to succeed in resurrecting. When this happens the artist will already be resurrected and will already have been added by the God to himself in the burning bush. When Picasso meets (finds) Marie Therese, the praxis of art will simultaneously begin to mutate, and the abandonment of Olga will be matched by the abandonment on behalf of the Painter of ancient Mannerism: now, a number of young girls may again find their body running on the beach, although according to the melodic figure proper to the score that animates their vital drive. Pan entered their veins up to the surface level of their skin like a tattoo. Later this figure will constitute the presentation in itself at surface level of the petrifaction in progress. At the end of the way of incarnation we have at the same time petrifaction and the ascent of Echo together with her encounter with Bacchus: the Goddess that petrifies on the beach can therefore give rise to a new conception in the abstraction of herself. Hence the new Lord of the garlands that will come to walk along the path of the Work. When Picasso makes his conversion to Classicism this is the path that will bring him into contact through the ongoing petrifaction with the reality of a Goddess (of meaning) who designs and fixes her own elegance on a mathematical level with reference to a specific nesting process. If God affects giving new life, the Goddess perfects and closes the Form in its truth. The hand of the artist who works out the extroversion will come to be acted upon by the God who manages to operate as new categorial in action. On the other side, the hand of the hero who works for his closure in the new conception comes to branch inside until it closes every possible space. Here is Picasso's Goddess on the beach identified by her points of maximum information. Hence her immaculate purity, a purity that cannot be distorted or manipulated under penalty of its destruction. Only if she is pure will she be able to conceive, only if she presents herself as a new axiomatic system but open in itself in its interiority. The ramification in progress discloses the reality of new semantic horizons but at the same time draws the nesting in itself of a system-soul. Here is the hand of an artist (G. B. Tiepolo) who recounts the branching by the Goddess (Daphne) of which he himself stands as a support. Hence the consequent coming to rise of new garlands of thought: the conceived can only interpret and support the way to abstraction. If the culmination point for the emerging Minotaur is given by the petrifaction followed by the ascent of Echo, the culmination point, instead, for the new conceived will be given by the *experimentum* and the irruption. This is the case with Vermeer. When the observer joins the Temple this means that there is the possibility of new conception at the level of meaning in action. Daphne that branches closing herself in her thoughts and nesting in the *Sylva* can really conceive while becoming in herself impenetrable. Here is an observer (Endymion) who can, therefore, close himself in his reflection in a cyclical eternity. In Mola we don't have the imperious Muse but the sweet companion, not the face of emotion but that of the rule in action. In this context the hand of Rodin is the hand of that God who to the extent that he self-organizes together with his meaning gives rise to a Work but in the truth, a God who at the same time presents himself with the face of the Muse that bursts out giving the life (here is the action of painting).. The sculptor forges his own neural circuits to express a work capable of infusing intentionality in the body of the new arising observer thus determining the expression of new life. But next to Rodin and Picasso there is Mola whose goal is not to build a work that can give life but rather to determine his own addition as observer to the Temple. Now there is no Muse (Clio) that bursts out, but Endymion that lurks in meaning and performs a phantasmal life. Here is a hero who closes himself in the reflection of himself, that is to say in a meaning that appears to be given for eternity. Endymion is the observer who actually adds himself to the Temple in view of the identification of the cyclic eternity of a body, his body. While Marsyas who comes to be added as creator ensures the elevation of his soul, Endymion ensures the eternity of a body, he exalts the rule in its purity not opening, therefore, to the way to abstraction and the *experimentum*. Here is an observer who does not simply drown in the image, but acquires through the image and its beauty an aesthetic vision of himself such as that exalted by Proust, thus drawing on an aesthetic dimension that appears able to ensure an eternal return. Closing himself in meaning and in the pure aesthetic dimension relative to his truth, he promotes the permanence and the eternity of his own body. In other words, he conquers an invariance that concerns only truth and not creativity. Mola in achieving this rarefied aesthetic dimension identifies himself with the Goddess as rule, enclosing himself in the Goddess and observing himself through his coming to reflect himself in her. On the one hand there is the God of emotion who imperiously invades (with Clio) the hero, on the other the Goddess that with infinite and congealed sweetness encloses in the same light of the Moon the hero who thus reaches his identification with his own body but on the basis of an eternal and cyclical return. Here is a form of eternal possession of one's own body, a body that cannot be denied because it is the repository of its own truth on an aesthetic level thus refusing to know dissipation and death. Truth of feeling and action of life. Here is the reality of an impenetrable night identified as *Sylva* that grows on itself: it shows the face of an observer, my face as well as the face of all the other heroes who, as observers, join the Temple by sacrificing to the Goddess. But here, on the other hand, is also a new and sunny day that turns a furtive glance at those ancient lifeless remains that testify to the sacrifice that has taken place. Body rendered to inertia and lifeless, on one side, and body rendered, on the other side, to an eternal return without, however, having ever known the action of the real life. A body without metamorphosis. Here is a stone that speaks of itself as eigenform and as *Via*. Starting from this stone, it is a new form of reflection that can come to life now, a reflection that will correspond to the emergence of a new kind of body: a body of simulation. Now the self-organizing Goddess comes to see (and reflect) her face but in a surreal dimension: the conception will no longer concern only Bacchus in his ancient garments but also unheard heroes, inhabitants of strange lands. Quaking ducks will be to sail the new thinking that is to be born. Monsters and artificial beings will come to fly in the skies. Hence the explosion of fantastic and sometimes appalling worlds from which unprecedented and asexual angels will come to stem. Now D. Marr, bringing his surrealist soul into a dowry, will be able to help the Painter to explore the depths that come to open up at the level of a renewed flesh and lead him to emerge from the spell of Marie Therese in view of a new synthesis, i.e. of that reversal of the praxis of art which will come to surface in Guernica. The ancient Gods rise and cry out their pain in the face of a reality that appears to come to devour them from within, a pain which is that of all humanity. It is Picasso, now, who must provide the score and later, together with Jacqueline, the same key to reading. The book will no longer be made of stone but of pure fire, the burning fire that animates Jaqueline's eyes, as related to the silent passion of a soul. In the act of creation, Picasso, as before him had already done Rodin, stands as the hand of the God, the hand that makes visible the invisible through the Method thus coming to shape the evolving mind of a new observer. Here is the new Minotaur that rises in Picasso's painting and through which the Painter becomes aware of himself and his destiny.

The brain comes to see by means of computations referred to fractals etc., computations which in the meantime they express themselves in accordance with the selection operated by the Method bring into play a Nature as a Work of art. Here is a geometry that will come to coincide with the very creation of new figures of a possible connection characterized by completely innovative mathematical modules. We will see a new world to the extent that we will be able to think of it in harmony both with the leavening of different architectures of our possible invention and with our coming to be selected by the Method of the God at the level of our cognitive development. In this way, at the level of the visual cortex, a (neural) geometry of the connections is created by means of a self-organization process: a neurogeometry that will be to weave itself on the basis of the nourishment offered by the original web of the programs. The reflection-mirroring in the countenance of meaning in action serves to enact a still more profound entrenchment. The cell closes on itself to ensure stability and autonomy, but on the basis of its own growth and self-identification, result will be both selection and evolution, invention and rediscovery. At cell level the "epistemological" work and the ongoing reflection play a vital role. At the level of life (and cognition) function cannot but self-organize together with its meaning. The cell's achieved autonomy must, then, necessarily refer to the gradual identifying of an "internal model" marking off (and scanning) within itself the process of selection taking place. In this autonomy we can recognize the early dawn of a consciousness of itself (but mathematically implemented) on behalf of the self-organizing system.

4.3 Shannonian Information Versus Instructional Information

As Carsetti remarks [16], at the level of natural evolution we have to distinguish a static stability related to a pure replication of the existing reality and a dynamic stability that keeps itself along the ongoing development but that is tied to continuous processes of transformation and innovation. At the level of this last type of stability, a biological organism can achieve a progressive increase of its inner complexity (thus determining an effective enrichment of its informational assets), only through the creative utilization of random shifts able to pave the way with a view to achieve higher (and initially unforeseeable) stages of stable complexity. It is through this creative procedure that the random game of evolution is coming to launch forward the growth process, allowing the emergence of new *equilibria*, of new structures capable of serving as genuine accumulation points for new leaps, for sudden variations in the growth rates of internal complexity. In this way such variations can come to determine the overcoming of ancient thresholds with a view to prime new levels of invariant stability. In this sense, we have to distinguish surface invariance from depth invariance, an invariance this last linked both to specific processes of growth and sudden openings up from the inside. With regard to this type of invariance, there is no longer simply the need to ensure both the fidelity of replication with respect to the original message and the permanence of mutations that have inserted themselves within the fabric of the message and which have proven to bear important selective advantages. What first appears necessary to ensure it is that a given structure may come to replicate itself providing both a systematic elimination of noise that cannot be utilized constructively and a real unfolding of non pre-determined openings up leading to the 'irruption' of creative noise thus ensuring a subsequent widening of the semantic foundations of the message to be transmitted, i.e. of the biological structure that replicates itself.

Thus, at the evolutionary level, we are faced with a dynamic situation that transforms itself through the achievement of successive *equilibria.* On the one hand, to preserve stability in the replication the coupled system must operate a continuous elimination of a variety of noisy pathways. This expulsion, from the island of the identical replication of underlying "openings up" and possible tension factors (as well as of random external aggressions), constitutes an objective impoverishment of the expressive capacities of the structure that reproduces itself as regards its own growth and its revelation on the deep level. On the other hand, the creative utilization of random shifts, the ordered inclusion of mutations etc. allow for a real growth of the different structures at stake but only if new possible horizons of stability in compliance with an adequate growth are identified at the mathematical and modeling level. In any case this creative utilization cannot be increased beyond a certain threshold if one does not wish the structure to lose its own identity, its own characteristics of internal stability and balanced control with respect to the ongoing disclosure processes.

It is exactly in the folds of this intrinsic dialectic that, in many respects, lies the origin of the metamorphoses of Nature, the reason or more precisely one of the reasons that drive the continuous emergence of new forms of life. From a real point of view, the issue is not, therefore, to clarify by what threshold, in the presence of mutations, can the invariant replication of a given structure be assured. The problem, rather, is that of explaining how and in what terms a conservative growth of the structure (able to disclose the deep "fabric" of the potentialities that underlie it), can simultaneously achieve both a creative use of noise and an adequate level of surface invariance, in such a way as to allow a coordinated alternation of periods of convergence and divergence. It is precisely the existence of this Janus' door with respect to Chance that alone may ensure, in accordance with the hypothesis outlined here, that launch forward of evolution that we have mentioned above. It is no longer, then, a matter of determining only the boundaries of a particular threshold, but of understanding how this threshold, while operating, could be moved forward, in compliance with specific connective modules and in accordance with a gradual increase in the levels of complexity.

The size relative to a process of self-organization, is that particular dimension in which the change of organizational modules is not governed by a predetermined program, but rather by a program that is produced by the encounter between the progressive realization of specific virtualities living within the evolving system, on the one hand, and, on the other hand, the revelation (as a result of the action of particular driving factors) of specific orderings articulating at the semantic level. This process may result, in principle, under the influence of a variety of factors, in a progressive reduction in the conditions of possible high redundancy proper to the initial state, and in a related and subsequent increase in potential variability at the symbolic level. This will allow a subsequent enlargement of the scope of action of the internal regulatory factors, an enlargement linked to the emergence of new constraints and renewed forms of organization. In this way, an increase in variability will come to correspond to an increase in specificity, to a lessening of disorder. And this without falling into paradoxes or contradictions. It is exactly this near-contextual increase in variability and specificity, which we refer to when we speak of an "opening up" of a coupled system endowed with self-organizing modules. In other words, when specific threshold are crossed in an "intelligent" way, the bases of variability may grow up and, at the same time, the conditions for realizing highly complex organization phenomena begin to be implemented.

The key point is represented by the fact that biological complex organisms are determined by a set of constraints, codes and programs but in the conditions, *in primis*, of real observation. Information that lurks in the deep levels of the different organisms specifies at the same time the structure of these same organisms. Following Layzer, we must, therefore, individuate, at first, a state space articulating at the biological level and, consequently, define the involved microstates or elementary events as they are determined by regulatory constraints, production rules, limit points etc. Variability and specificity should, therefore, be defined in relation to this particular type of microstates: in relation, that is, to elementary events that cannot be identified through a naive reference to frequencies or surface constraints but which must, on the contrary, be defined, on an objective ground, in the framework of the dialectic relationship that exists between the observer and the source, between hypotheses and falsifications, with reference, above all, to the very possibility of triggering, through the guidance offered by the measurements and observation procedures, a process of freeing up of depth information and, therefore, the progressive revelation of the deep structures linked to this process. Here we can find one of the crucial nodes of the ongoing analysis. Our ability to come to distinguish particular microstates able to account for the articulation of an entropy-function linked to specific regulatory constraints can allow us to draw a first outline of the threshold and self-organisation processes that are at the basis of that particular kind of living structure represented by the DNA but only to the extent that we have been able to put in place adequate observation tools in order to allow a real growth of the evolving entire coupled system. In this sense, DNA necessarily appears as a molecule that has at its core the rules responsible for the full expression of its programme as well as the rules that are indispensable in view of changing these same rules.

In other words, it is necessary first of all to realize that we cannot calculate biological information the way we do in the case of the transmission of signals. We cannot confuse macroscopic events or macrostates with microstates. Nor is it sufficient to take account of the intervention of the measuring activity, distinguishing free information from bound information, when we are unable to locate the intrinsic reality of biological microstates and, therefore, the 'type' of the specific constraints linked to them. It is necessary, instead, to individuate the levels of depth information, where the regulatory constraints are hidden; it is furthermore necessary to account for the relationship joining the observer to the source, and in particular, to explain the link between the different levels in which the information content is spread. This will allow us to define properly the microstates with reference to the real evolution of the source and the progressive appearance of new constraints (but in the presence of an increase in variability). As Jaynes rightly noted, going into the depth levels of the source, in this case of the DNA, is only possible with the help of very sophisticated telescopes/models, with the help, in particular, of measures of information and hypotheses that are not rigidly predetermined. Measures and hypotheses, we must add, that should be capable of explaining the complex intertwining that exists between surface information and depth information as well as of accounting for how noise, generated primarily at the internal level, can then be utilised creatively thus becoming a driver of innovation. All this means that the microstates cannot be considered as mere letters of an alphabet, as entities anchored to one-dimensional surface constraints, unable to explain the dialectics between source and observer. At the same time, a growth in creativity on the part of the system cannot arise by means of simple additions of new states-letters, it may come to bloom only in dependence of global reorganisation processes that occur through the internal splitting of the molecules, the appearance of new fixed points, the identification of new attractors etc. The real problem is to figure out how the auto-poietic (non-trivial) machine may come to widen, autonomously, the basis of its internal variability in order to lay the foundations for the realisation of a historically determined evolutionary tinkering, that winds through a passage from one element to another in a hierarchy of levels, according to modules that cannot be entirely predetermined.

Indeed, as Atlan and Carsetti correctly remark [17], in a natural self organizing system (a biological one, characterized by the existence of cognitive activities) the goal has not been set from the outside. What is self-organizing is the function itself with its meaning. The origin of meaning in the organization of the system is an emergent property. In this sense the origin of meaning is closely connected to precise linguistic and logical operations and to well defined procedures of observation and self-observation. These operations and procedures induce, normally, continuous processes of inner reorganization at the level of the system. The behaviour of the net, in other words, possesses a meaning not only to the extent that it will result autonomous and to the extent that inside it we can inspect a continuous revelation of hidden properties, but also to the extent that it will result capable of observation and self-observation as well as intentionally linked to a continuous production of possible new interpretation acts and reorganization schemes. The state space relative to these observational, self-observational and intentional functional activities cannot be confined, in this sense, on logical grounds, only within the boundaries of a nested articulation of propositional Boolean frames. On the contrary, we will be obliged, in order to give an explanation, within a general theory of cellular nets, for such complex phenomena, to resort to particular informational tools that reveal themselves as essentially linked to the articulation of predicative and higher-order level languages, to the outlining of a multidimensional information theory and to the definition of an adequate intensional and second-order semantics.

Here is the first outline of the concept of self-organizing model. The design of the afore mentioned new kind of semantics, if successful, will necessarily conduct us to perceive the possibility of outlining a new and more powerful theory of cellular automata, of non-trivial automata in particular, that will manifest themselves as coupled models of creative and functional processes. We shall no longer be only in the presence of classification systems or associative memories or simple self-organizing nets. We shall be, on the contrary, faced with a possible modeling of precise biological activities, with biochemical networks or biochemical simulation automata capable of self-organizing, as coupled systems, their emergent behaviour including their same simulation functions. When we consider, for instance, DNA as a complex system characterized by the existence of a precise language articulating within the contours of a self-organizing and "intentional" landscape, we are necessarily faced with a molecular semantics that needs, in order to be understood, explanatory tools much more powerful than those provided by the traditional Information Theory and classical Cybernetics. A particular embodiment of these tools, can be actually, represented by the outlining of simulation automata able to prime new forms of conceptual "reading" of the information content hidden in the text provided by the molecular language.

The real problem is to follow the living "contours" of the evolution of deep generative dynamics, in order to be, finally, able to prime, in a coherent way, the unfolding of depth information. But this is, precisely, one of the primary reasons that have led to the actual outlining of simulation models at the level of biological sciences: to create, that is, the conditions for the revelation of the deep information content of the source: to become a sort of arch and gridiron for the construction and the recovery of the Other (the Source, the original life) through the constraints relative to the successful realization of the paths of simulation. Thus, when we speak, within this frame of reference, of molecular semantics and, in perspective, of neural semantics, we actually recognize, first of all, the existence of functional principles of self-organization that articulate at the deep level of the symbolic and generative dynamics according to linguistic patterns of construction. In this sense, for instance, if we take into consideration particular dynamic Halmos algebras in order to model, at the predicative level, a specific database and if we want to arrange this data base as a self-organizing and intensional net, as a sort of "intelligent" neural network, we have to delineate, besides the syntactic architecture of the entire net, also its intensional inner dimension. In particular, we have to define the patterns of the information flow, the roles played by the semantic constraints, the action expressed over time by the different meanings and so on. The resulting, semantic model doesn't constitute, thus, only a form of abstract representation or a simple frame of neural computations: it will present itself, on the contrary, as a real simulation tool capable of establishing the necessary conditions for the "opening" of a linguistic dialogue (of a coupled process of information production) with the deep (and self-organizing) information content as it articulates, for instance, at the level of the human mind.

For many years now even the great American scholar S. Kauffman came confronting the issues concerning semantic information to the point of advancing in a recent paper a significant proposal for the delineation of a Physics of Semantics (or rather a molecular Semantics) thus going beyond the traditional information theory as developed by Shannon, a theory this latter based still on a type of mathematics that is too simple and therefore incompatible, in his opinion, with the complexity of vital phenomena. "Shannon information—he writes—require that a prestated probability distribution (frequency interpreted) be well stated concerning the message ensemble, from which its entropy can be computed. But if Darwinian preadaptations cannot be prestated, then the entropy calculation cannot be carried out ahead of time with respect to the distribution of features of organisms in the biosphere this, we believe, is a sufficient condition to state that Shannon information does not describe the information content in the evolution of the biosphere. There are further difficulties with Shannon information and the evolving biosphere. What might constitute the "Source"? Start at the origin of life, or the last common ancestor. What is the source of something like "messages" that are being transmitted in the process of evolution from that Source? The answer is entirely unclear. Further, what is the transmission channel? Contemporary terrestrial life is based on DNA, RNA, and proteins via the genetic code. It is insufficient to state that the channel is the transmission of DNA from one generation to the next. Instead one would have to say that the actual "channel" involves successive life cycles of whole organisms. For sexual organisms this involves the generation of the zygote, the development of the adult from that zygote the pairing of that adult with a mate, and a further life cycle. Hence, part of one answer to what the «channel» might be is that the fertilized egg is a channel with the Shannon information to yield the subsequent adult. But it has turned out that even if all orientations of all molecules in the zygote were utilized there is not enough information capacity to store the information to yield the adult. This move was countered by noting that, if anything, development is rather more like an algorithm than an information channel [...]. In short, a channel to transmit Shannon information along life cycles does not exist, so again Shannon information does not seem to apply to the biosphere. It seems central to point out that the evolution of the biosphere is not the transmission of information down some channel from some source, but rather the persistent on going, co-construction, via propagating organization, heritable variation and natural selection, of the collective biosphere. Propagating organization requires work. It is important to note that Shannon ignored the work requirements to transmit "abstract" information, although it might be argued that the concept of constraints is implicit in the restrictions on the messages at the Source. While we mention this, we have no clear understanding physically of what such constraints are" [18]. It is in reference to such a conceptual framework that Kauffman comes to revisit Schrödinger. "We believe Schrödinger was deeply correct, and that the proper and deep understanding of his intuition is precisely that an a-periodic solid crystal can contain a wide variety of micro-constraints, or micro-boundary conditions, that help cause a wide variety of different specific events to happen in the cell or organism. Therefore we starkly identify information, which we here call "instructional information" or "biotic information", not with Shannon, but with constraints or boundary conditions. The amount of information will be related to the diversity of constraints and the diversity of processes that they can partially cause to occur. By taking this step, we embed the concept of information in the ongoing processes of the biosphere for they are causally relevant to that which happens in the unfolding of the biosphere. We therefore conclude that constraints are information and, as we argue below, information is constraints which we term as instructional or biotic information to distinguish it from Shannon information. We use the term "instructional information" because of the instructional function this information performs and we sometimes call it "biotic information" because this is the domain it acts in, as opposed to human telecommunication or computer information systems where Shannon information operates. This step, identifying information as constraints or boundary condition, is perhaps the central step in our analysis. We believe it applies in the unfolding biosphere and the evolving universe, expanding and cooling and breaking symmetries, that we will discuss below. Is this interpretation right? It certainly seems right. Precisely what the DNA molecule, an a-periodic solid, does, is to "specify" via the heterogeneity of its structural constraints on the behavior of RNA polymerase, the transcription of DNA into messenger RNA. Importantly this constitutes the copying or propagating of information. Also, importantly, typically, the information contained in aperiodic solids requires complex solids, i.e., molecules, whose construction requires the linking of spontaneous and non-spontaneous, exergonic and endergonic, processes. These linkages are part of the work cycles that cells carry out as they propagate organization. [...] The working of a cell is, in part, a complex web of constraints, or boundary conditions, which partially direct or cause the evens which happen. Importantly, the propagating organization in the cell is the structural union of constraints as instructional information, the constrained release of energy as work, the use of work in the construction of copies of information, the use of work in the construction of other structures, and the construction of further constraints as instructional information. This instructional information further constraints the further release of energy in diverse specific ways, all of which propagates organization of process that completes a closure of tasks whereby the cell reproduces" [19]. However, in the opinion of the great scholar, it is necessary to define the contours of this instructional information: "Just as Shannon defined information in such a way a to understand the engineering of telecommunication channels, our definition of instructional or biotic information best describes the interaction and evolution of biological systems and the propagation of organization. Information is a tool and as such it comes in different forms. We therefore would like to suggest that information is not an invariant but rather a quantity that is relative to the environment in which it operates. It is also the case that the information in a system or structure is not an intrinsic property of that system or structure; rather it is sensitive to history and environment. [...] Information is about material things and furthermore is instantiated in material things but is not material itself. Information is an abstraction we use to describe the behaviour of material things and often is thought as something that controls, in the cybernetic sense, material things. So what do we mean when we say the constraints are information and information is constraints [...]" [20]. "The constraints are information" is a way to describe the limits on the behaviour of an autonomous agent who acts on its own behalf but is nevertheless constrained by the internal logic that allows it to propagate its organization. This is consistent with Hayle's [...] description of the way information is regarded by information science: "It constructs information as the site of mastery and control over the material world". She claims, and we concur, that information science treats information as separate from the material base in which it is instantiated. This suggests that there is nothing intrinsic about information but rather it is merely a description of or a metaphor for the complex patterns of behaviour of material things. In fact, the key is to what degree information is a completely vivid description of the objects in question" [21].

"The first is that the number of possible messages is not finite because we are not able to prestate all possible preadaptations from which a particular message can be selected and therefore the Shannon measure breaks down. Another problem is that for Shannon the semantics or meaning of the message does not matter, whereas in biology the opposite is true. Biotic agents have purpose and hence meaning. The third problem is that Shannon information is defined independent of the medium of its instantiation. This independence of the medium is at the heart of a strong AI approach in which it is claimed that human intelligence does not require a wet computer, the brain, to operate but can be instantiated onto a silicon-based computer. In the biosphere, however, one cannot separate the information from the material in which it is instantiated. The DNA is not a sign for something else it is the actual thing in itself, which regulates other genes, generates messenger RNA, which in turn control the production of proteins. Information on a computer or a telecommunication device can slide from one computer or device to another and then via a printer to paper and not really change, McLuhan's "the medium is the message" aside. This is not true of living thing. The same genotype does not always produce the same phenotype" [20]. And later the great American scholar adds: "According to the Shannon definition of information, a structured set of numbers like the set of even numbers has less information than a set of random numbers because one can predict the sequence of even numbers. By this argument, a random soup of organic chemicals has more information that a structured biotic agent. The biotic agent has more meaning than the soup, however. The living organism with more structure and more organization has less Shannon information. This is counterintuitive to a biologist's understanding of a living organism. We therefore conclude that the use of Shannon information to describe a biotic system would not be valid. Shannon information for a biotic system is simply a category error. A living organism has meaning because it is an autonomous agent acting on its own behalf. A random soup of organic chemicals has no meaning and no organization. We may therefore conclude that a central feature of life is organization—organization that propagates" [21].

4.4 Omega World: Incompressibility and Purity

When the Silenus feels again the hand of the God to act at the level of his interiority it is because he was added, but the God who comes to speak in him can do so only in the emergence of a new incarnation. Here we can find a theoretical limit as regards the main theses put forward by Chaitin: extroversion is not carried out in view of a simple optimization of the conditions of the ongoing self-organization, but to undergo that selection which can open up to a renewed incarnation, an incarnation that alone can determine the expression of a real novelty. It is in this way that I can place myself as a stool for the God but in my own metamorphosis. The new incarnation in the resurrection is, therefore, in function of the novelty but following the achievement of the recovery. By performing the extroversion with reference to the ongoing design of an adequate recovery, the conceived opens up to the intervention by the God who manages to impress the signs of his renewed creativity. He sinks his knife into the flesh of the conceived within the framework of a lunar atmosphere. Marsyas burns in the experimentum thus finally coming to see the realization of his elevation to the soul. The hand of the God who shakes the artist in his roots is the hand that will come to sanction the Work but in what is the renewal by the same God of his creativity: here is the conceived that sees the revealing to him of the God (in his coming to be added by the God to himself) through the action of painting as aroused in him by Clio: he will stand as the last witness of this same action. It is in this moment that in Vermeer almost for the first time the Painter enters the scene showing, in particular, on himself the marks left by the selection in progress. Here is the true legacy and here is the outline of an adequate set of eigenvalues, those eigenvalues, for instance, that will come to populate Picasso's painting "The flute of Pan" coming to emerge on the very skin of the arising Minotaur along the course of the incarnation like a real tattoo. At the level of the praxis of art, we must therefore distinguish, for instance, between the hand through which P. F. Mola provides to tell and identify himself in the reflection of an invariant eternity and the hand through which Vermeer comes to burn in the bush but according to a metamorphosis and the onset of new creativity. Hence the nesting of omega in itself and in its own abyss in accordance with the intuitions pursued by L. Fontana in his exploration of the "hole". The conceived burns on the gridiron but providing for the release of what is his true inheritance: what you really love remains. It is a new praxis of art that now comes to incarnate in me (as an emerging Minotaur) and that presents itself with the face of the Muse. The Muse represents the same revelation of this praxis, the moment of my death as Silenus and of my resurrection in the metamorphosis, the very cry of a new mind that captures me in its inspiration. I see my inheritance while I come to be lived by it (the tattoo) exactly in the moment, that is to say, in which I come to stand like the risen one, like the new revived Painter who provides to tell his own metamorphosis (Melville). He who resurrects can only look at his ancient remains. Telling the metamorphosis means witnessing one's own resurrection but entrusted to a Work, it means realizing this same Work but in the whirl of emerging worlds. The novelty, in any case, does not guarantee compensation. The story coincides with recovery and adjunction. But the story is the Work and the Work is History: the adjunction of the hero and the fulfillment of the Work are closely united and coincide with the discovery of new creativity and the end of the self-abandonment by the God but on my skin of Silenus (and of shipwrecked). Here is a creativity fit for a story and a legacy and here is a metamorphosis continuously in action. Hence the birth of new possible horizons of optimization with respect to the ongoing organization. The one who tells is already the Other and can only be glimpsed in backlight and along the descent on his part into the vertigo of a new purity (Fontana) which also presents itself as new incompressibility. This is the path that opens up to the new outcrop along the same coming to light up of a new categorial (with new flames as new concepts in action). Here is the praxis that will come, therefore, to preside over the new adventure of the Minotaur, of he who comes to be born from the bed of intensities and who will come to be acted by the new concepts in action. Hence new Nature and new outcrop but in dependence on the Work of art created and the selection made. In this way we are faced both with a Goddess that comes to speak to us to the extent of a meaning that self-organizes together with his creativity and with a God who in the outcropping will come to express himself, in turn, as a new function that self-organizes together with its meaning: Nature in dependence on the Work and vice versa. Chaitin does not identify the role played by the Muse and does not focus on the centrality of the transition to the new incarnation. He, however, is perfectly aware that the outcrop is only possible as a result of extroversion and disembodiment as well as of the recourse to specific metamathematical procedures. Without such recourse there is, in effect, no possible selection nor, therefore, authentic novelty. Here is the revolutionary intuition related to the recourse to the software space and to the methods of metamathematics.

Hence the new design of the relationship between the God and the humans as creators who through the ring-threading animate the variegated universe of the Work. Once we will have observers who with their visions will intersect Nature and once we will have creators who with their thoughtful garlands will come to ring the Work (Opera): here is Marsyas as Lord of the garlands but here is also the God with his knife. The hand, however, that traces the emotions that connect the Muse to the hero who suffers the *experimentum* is the hand of that same hero, the hand that "sings" the emotions of the hero along his own journey through the realization of the Work (Opera). The dying hero rises to poetry and its eternity. It is his soul that he comes to discover and that allows him to prepare for the new incarnation. He who draws his own metamorphosis comes to feel the God speaking to him: the invisible becomes visible for him and in him. Here is Clio who reveals herself but in the moment in which she comes to emerge in the new irruption acquiring self-awareness (but in the Other). Emotion overwhelms the Muse illuminating her in all her clarity: truth and History combined in themselves. Here is the hand of a hero who has already been added along his own coming to die in the metamorphosis.

As we have just said, it is the hand of the Silenus that works for the recovery thus opening up to that new categorial that will act from within in order to come to shape my embodiment as emerging Minotaur. Here, then, is a revelation that can be given only through an incarnation, the constitution of a renewed imagination and the realization of a new stumbling block. The Muse does not limit herself to singing the deeds and the works, she will come to dress the same works of the Painter (cf. Jacqueline wearing Picasso), she will guide him into the realm of the hallucinogens, she will give herself to him by making herself part of him. Hence the selection operated by the hand of the God. This is the path that will allow the emerging of a new creativity, a creativity subtended, of necessity, by an adjunction. Here is the key role played by the conceived as Lord of the garlands (Carsetti 2004). Creation necessarily passes through the *experimentum*, that *experimentum* that only the divine child can come to face. This is what the offer of the stool means: the overcoming, that is to say, of the "lack" (relative to the self-abandonment) on the part of the God with respect to himself. If the God does not suffer abandonment he will not be able to know-generate authentic novelty. But use can only determine the abandonment, an abandonment that refers primarily to the closure and the petrifaction that necessarily follows the surfacing by the God. Hence a new possible conception. Endymion stops at the Pillars of Hercules and joins the Temple. In the "action of painting" by Vermeer, on the contrary, the Painter goes beyond the Columns and comes to explore the related abysses through the thoughtful use of his brush. The Painter tells of a malware that affects the God, a malware situated between absence, devouring and self-abandonment. The hero is therefore called to run to the rescue of the God, to enter, that is to say, the fields of a new Semantics but in the footsteps of a radical metamorphosis: it is a new incompressibility that opens up before his eyes and with it the need for new purity.

In such a framework the reality of the outcrop represents the face of the God unfolded in himself in his own concept, a concept that becomes spatial (Fontana) that is pure space in itself absolute and divine as regards its truth as well as what is its only possible colour: the gold colour relative to the vision in the truth on the part of the God. It is this outcrop that comes to be torn by the awl that sinks at the level of the surface, opening up a new, further hidden dimension relative to the "hole". It is the canvas that comes to be torn and at the same time the veil: beyond the reality relative to the unfolding taking place at surface level, a further imagination space thus comes to delineate itself in depth: eigenvalues in place of surface eigenforms. Hence a different temporality, a different role for meaning; hence the dialectic itself relative to the observer (and his self-reference operations), to the second-order structures and to the opening up to new Semantics. But from here also stems the possibility of the extroversion and the intervention of the hand of the God. Now the penis of the divine child that is without sin and which the Painter was authorized (at the level, for example, of the great Florentine and Venetian painting) to show on the surface in its nakedness as the only Way to salvation appears, in fact, reversed inside in its depth. We, now, find the incompressibility and purity related to the penis in the "hole" which comes to be articulated according to a retraction. Here is a mysterious dimension that opens up, the dimension relative to a void, to a possible abyss that in its turn opens up, to a real incompressibility that can only be thought of or brought back to an experience of pure mathematical thought. A dimension to which, for instance, the omega number of Chaitin opens, once properly defined and used. We cannot, however, limit ourselves to preaching and witnessing the active presence of a void, but must, in fact, face the journey: here are the ancients Egyptians and the delineation of the Road in the depths of Hades, but here also is the path within the realm of non-standard mathematics, a mathematics related to imaginary, non-standard, surreal numbers, etc. Here is the fundamental relationship between Work and Nature. Fontana traces the penis as a negative image relative to the abyss that concerns its purity and obtains a first presentation of an infinitistic method in action. He sees the canvas which, once torn, starts a metamorphosis of a temporal and quantum character. He rediscovers the value of saving purity but by bringing it back to the level of openness and the gash, of the change of semantics, and of the exploration of the realm of undecidability with the growth and affirmation of new number systems. To this growth, however, will come to correspond new types of rooting of the senses, new vision and new natural processes, the same material and unprecedented reality relative to a new praxis of art (as happens in the late Titian).

Once the veil has been torn off and we have entered vertigo, the first thing we discover is the presence of an interconnected temporal dynamics linked to a variant fabrics of eigenvalues, a dynamics that attacks the very evolution of our sensibility from within and through which it is transformed. To exist means, now, to reconquer oneself each time, to be reborn but in the identification of one's inheritance, to displace oneself in the metamorphosis. Only a dialectical thought can come to our rescue: we need, for instance, a close government on the part of the brain of the always mysterious relationship between sensibility and intellect. Fontana with his awl actually opens to the dimension of complex numbers, and refers to mysterious sets of eigenvalues and a congealed temporality. He enters the complexity of the journey of a soul and succeeds, through his extensive meditation on the crucifixion, in extracting pieces of flesh that have turned into lapis lazuli that hide the secrets of a thought that can again find itself only in correspondence with an emotion that regenerates it.

When Picasso in the painting: "The flute of Pan" paints himself as an emerging Minotaur (which rises from the remains-inheritance of Marsyas) he is able to do so because as a painter he has been added to himself by the God, a God who can, therefore, come to free-deploy his nature in its classical sense: hence a recovered Nature in action. It is because the Painter was added and because the God comes to see (and create) in and through him that Picasso can trace and paint his story by identifying the stages of what was his own conception. When he identifies himself in the painting as emerging Minotaur, it is actually because the God has already come to see through him: the evolution-growth of the Minotaur is coeval with the metamorphosis by the God in a Nature intersected by perceptual acts (and observers). This means that the Painter must have come out of Mannerism (as well as from the world of Olga) and must have been added in accordance with the completion of that key painting: "The Crucifixion" which anticipates, in some respects, the conclusion of his late Mannerist phase. His adjunction shows itself, on the operational level, precisely through his passage to Classicism, his very coming to trace, for instance, his self-portrait according to the classic countenance typical of a painter who has proved himself able to achieve recovery. Meanwhile, however, the Painter will have suffered consumption and death, he will have allowed the God to get out of self-abandoning and will have experienced that same abandonment even through the abandonment of his wife Olga on his part. He must come to be reborn to (and in) the praxis of art: only in this way can he come to find in himself the dimension proper to Classicism. Here is the journey of the Minotaur accompanied by his classic mentor, a journey that through petrifaction will lead to new conception. Here is a Painter who, to be able to rise again to creativity, will also have to face the difficult step represented by the achievement of recovery. Only if this step is successfully passed will the new Marsyas be able to open up to a new praxis of art. When the Painter is added he rises again in the Minotaur (which refers to the surfacing of the God) in accordance with the ongoing metamorphosis. Then the Minotaur gives rise to a new Marsyas-Painter who will make up his story in the abstraction in order to determine (and paint) his own extroversion and to be engraved by the God with reference to the recovery made. Hence the story of his own funeral on the part of the Painter (cf. Titian and Sironi) as a stopping term for metamorphosis: now there are only devout souls and, at the same time, the memory represented by the Painter's painting-testament.

Marsyas who is reborn and who narrates his story starting from the Minotaur that petrifies, is the Painter who makes the transition to Classicism but is also the Painter who comes to meet Marie Therese. He encounters on a natural and real level the woman who, as a Muse, observes and supports him, and who is also part of the surfacing taking place on the part of the God as Nature in itself classical. On the one hand, the Painter who is added allows the God to emerge from self-abandonment, on the other, he will then give rise to a new Marsyas-Painter to the extent that he has come to transform himself into a new Minotaur just as this last comes to express himself at the level of the outcrop. This is the Painter who will paint his encounter within the outcrop with a Muse in herself classical, an encounter which can only be realized on a natural (and real) level and which will allow him to experience the rediscovered radiance of the God. The Painter-Marsyas is real to the extent that he comes to recount his encounter with the Muse but the reality of the Muse will be sanctioned at the level of the Work and the praxis of art. Marie Therese is the painting that depicts her and lives in the God that unfolds: the painting that concerns her constitutes the reality that still holds her to herself. Here is Bluebeard's castle but here is also the Museum of the works, and a memory giving rise to new life expressions. In the film of Lars von Trier: "The House that Jack built", Jack can come and kill in pursuit of the ghost of that portrait which for him constitutes the true face of the soul. Picasso who depicts the crucifixion, works for the surfacing of the God in the very abandonment of his Mannerist creed and of his orderly life with Olga. He represents his own crucifixion at the merciless rhythm of a dance to the point of exploring the contours of a new break-in thereby opening up to a new phase of the praxis of art, a phase characterized by the presence of Marie Therese as a new Muse. He is now Marsyas as the new prince of simulation who paints at the level of the dimension proper to a dream which is, in fact, pure memory: here is his story, his path as Minotaur, his own origin as marked from precise eigenvalues, his very coming to emerge at the natural level thus revealing himself as new flesh. When the Painter completes his journey at the level of the Work, he actually created an extroversion in view of the *experimentum*. He is Titian who paints the rising of the Nymph-Minotaur but with reference to his own remains as Marsyas. He has proceeded to investigate himself as simulation in progress, as the one who in the bush experiences the conversion of the artificial into the natural. His art and his vision can only have a classical value: the God will come to live in him to the extent that he will have been added: in effect, the Painter will come to situate himself in the flesh to the extent of the completed adjunction. In this sense, the true reality will come to be represented by that adjunction in creativity which will lead him beyond the confines of Reflexivity allowing him to grasp the radiance that animates Marie Therese but in accordance with a dimension which is no longer simply physical (i. e. referring to the paradigm of naive physics) and that makes reference, on the contrary, to a new conception of the body, to a body understood as renewed incompressibility. Hence a creativity that reveals itself capable of self-reflecting but according to a process of self-organization. Here is the purity of the new conceived (of the new Lord of the garlands) as he always emerges from the depths of the abyss. Here is the role played by the entrance into the scene of new and unprecedented number systems as well as new types of orderings. Here is the Muse on the beach presenting herself as pure mathematical construction, as the dynamic balance proper to thought in action, that balance which will only allow us to identify in mathematical terms the new dimension of reality introduced by Classicism. It is, precisely, in correspondence with these conditions that the meeting with D. Marr comes to be realized: here is the ritual of the knife and the challenge with the Chance, here are the incisions and the blood, and here are the gloves, in themselves like a mysterious heirloom. From the ghosts of depression derives an ongoing self-abandonment by an innovative character that comes to hit Picasso in his first encounter with Dora. As for now, the realization of recovery will be different: the radiance of Marie Therese appears to be furrowed by the lines of an inner split that appears as genetic and at the same time universal. When Picasso comes to overcome the Marie Therese universe there will no longer be a Minotaur (with the legacy of Pan) who comes to meet Marie Therese as Muse, on the contrary, there will be the coming into being of a self-reflexive reality that presides over the discovery of a completely different and disturbing interior dynamics, a dynamics that Picasso comes to feel as close to his soul and that jeopardizes the certainties gained. Now there is no longer Bacchus who frees, the Bacchus, that is to say, who comes to marry Echo giving birth to the new conceived on the basis of a precise metamorphosis. Now there is a hero who comes to experience consumption and death but in view of new irruption. Now a new Muse will be born and will be represented in the case of Picasso precisely by Dora Marr. but then also by Jacqueline.

References

1. Dougherty, E. R., & Bittner, M. L. (2010). Causality, randomness, intelligibility, and the epistemology of the cell. *Current Genomics, 11*(4), 221–237. [p. 20].
2. Dougherty, E. R., & Bittner, M. L. (2010). Causality, randomness, intelligibility, and the epistemology of the cell. *Current Genomics, 11*(4), 221–237. [p. 21].
3. Dougherty, E. R., & Bittner, M. L. (2010). Causality, randomness, intelligibility, and the epistemology of the cell. *Current Genomics, 11*(4), 221–237. [p. 25].
4. Dougherty, E. R., & Bittner, M. L. (2010). Causality, randomness, intelligibility, and the epistemology of the cell. *Current Genomics, 11*(4), 221–237. [p. 26].
5. Dougherty, E. R., & Bittner, M. L. (2010). Causality, randomness, intelligibility, and the epistemology of the cell. *Current Genomics, 11*(4), 221–237. [p. 29].
6. Carsetti, A. (2000). Randomness, information and meaningful complexity: Some remarks about the emergence of biological structures. *La Nuova Critica, 36*, 47–109.
7. Bais, F. A., & Doyne Farmer, J. (2008). The physics of information. In P. Adrians & J. Van Benthem (Eds.), *Philosophy of information* (pp. 609–684). Amsterdam.
8. Carsetti, A. (2013). *Epistemic complexity and knowledge construction*, New York; Carsetti, A. (2000). Randomness, information and meaningful complexity: Some remarks about the emergence of biological structures. *La Nuova Critica, 36*, 47–109.
9. Wallace, R. (2014). Cognition and biology: Perspectives from information theory. *Cognitive Processing, 15*(1), 1–12. [p. 2].
10. Reeb, G. (1979). *L'analyse non-standard vieille de soixante ans?*. IRMA: Strasbourg.
11. Carsetti, A. (2019). Conceptual complexity and self-organization in life sciences. In *Complexity and integration in nature and society*. EASA: Salzburg.
12. Carsetti, A. (2013). *Epistemic complexity and knowledge construction*. New York.
13. Manzano, M. (1996). *Extensions of first-order logic* (p. XVI). Cambridge.
14. Németi, I. (1981). Non-standard dynamic logic. In D. Kozen (Ed.), *Logics of programs, lecture notes in computer science* [p. 131].
15. Henkin, L. (1950). Completeness in the theory of types. *Journal of Symbolic Logic, 15*, 81–91.
16. Carsetti, A. (2000). Randomness, information and meaningful complexity: Some Remarks about the emergence of biological structures. *La Nuova Critica, 36*, 47–109.
17. Atlan, H. (2000). Self-organizing networks: Weak, strong and intentional, the role of their underdetermination. In A. Carsetti (Ed.), *Functional models of cognition* (pp. 127–143). Kluwer A. P.: Dordrecht; Carsetti, A. (2000). Randomness, information and meaningful complexity: Some remarks about the emergence of biological structures. *La Nuova Critica, 36*, 47–109.
18. Kauffman, S. A., Logan, R. K., Este, R., Goebel, R., Hobill, G., & Shmulevich, I. (2008). Propagating organization: An enquiry. *Biology and Philosophy, 23*, 34–35. [pp. 34–35].
19. Kauffman, S. A., Logan, R. K., Este, R., Goebel, R., Hobill, G., & Shmulevich, I. (2008). Propagating organization: An enquiry. *Biology and Philosophy, 23*, 34–35. [p. 36].
20. Kauffman, S. A., Logan, R. K., Este, R., Goebel, R., Hobill, G., & Shmulevich, I. (2008). Propagating organization: An enquiry. *Biology and Philosophy, 23*, 34–35. [p. 39].
21. Kauffman, S. A., Logan, R. K., Este, R., Goebel, R., Hobill, G., & Shmulevich, I. (2008). Propagating organization: An enquiry. *Biology and Philosophy, 23*, 34–35. [p. 38].

Chapter 5
The Metamorphoses of the Revisable Thought and the Evolution of Life

Abstract Eurydice, living among shades, fails to "see": on the wave of the splitting she explores worlds conceptually, thinks *via* concepts, and delineates extensions (at the level of mathematical invention) in view of a possible new incarnation ensuing the new song which burst forth. With the birth of the Nymph new, specific forms of recursion and ordering will be possible, linked to a renewed many-sorted version, with new computational practices (at the mental level) which, however, will employ new methods and which will emerge from the chaos ensuing devastation, also in dependence of the surfacing of new language. The result will be the birth and consolidation of a new invariance, with new methods and new individuals at play. When, in fact, new fixed points will emerge at the co-evolutionary (and infinitary) level on the basis of new limitation procedures, we will witness new eyes in the flesh, and new Nature, as well as the actual establishment of new intended models, albeit in reference to more complex computational practices than hitherto. We will have been worn by new glasses consonant with more adequate systems of numbers, and will be able to find again the ancient natural numbers, and operate with them in accordance, however, with a more extensive savor.

5.1 Husserl and Goedel: Categorial Intuition and Meaning Clarification

Merleau Ponty, as is well known, is in line with Brentano and Koffka in considering the phenomenal *Umwelt* as 'already there', perception consisting precisely in detaching (*dégager*) the nucleus of this 'already there'. The distinctive nature of Gestalt is not as something alive in itself, independently of the subject which has to "insert" into it its relationship with the world; nor, however, is it constructed by the subject. It is not absolute, since experience shows that it can be made to vary, yet nor is it purely related to the Self, since it provides an *Umwelt* which is objective (transcendent). In this sense perception does not constitute a simple act of synthesis.

According to this viewpoint, Quine too considers, for instance, predication as something more than mere conjunction (the mere synthesis, Brentano would have said, of a subject-notion and a predicate-notion), not least since it ultimately coincides

A. Carsetti, *Metabiology*, Studies in Applied Philosophy, Epistemology and Rational Ethics 50, https://doi.org/10.1007/978-3-030-32718-7_5

with an act of perception. When we say ‘the apples are red’, this for Quine means that the apples are immersed in red. Predication indeed finds its basis on a far more complex act than simple conjunction-composition.

It should, however, be underlined that when in his later work Quine gives an illustration of the kind, he is quite consciously and carefully re-examining not only some of Brentano’s original ideas on the thematic of perception, but also a number of basic assumptions behind Husserl’s idea of relations between perception and thought. Can colour be grasped, Husserl asked, independently of the surface supporting it? Quite clearly not: it is impossible to separate color from space. If we allow our imagination to vary the object-color and we try to annul the predicate relative to the extension, we inevitably annul even the possibility of object-colour in itself, and reach an awareness of impossibility. This is what essence reveals: and it is precisely the procedure of variation which introduces us to the perception of essence. The object’s eidos is constituted by the invariant which remains unchanged throughout all the variations effected.

In Husserl’s opinion, together with perception it is necessary to conceive of acts based on sensory perceptions in parallel with the movements of categorial foundation taking place at the intentional level [1]. These acts offer a precise “fulfillment” to the complex meanings which for us constitute the effective guides to perception. When I observe gold, I see not yellow on the one hand and gold on the other, but ‘gold-is-yellow’. ‘Gold is yellow’ constitutes a fact of perception, i.e. of intuition. The copula, the categorial form par excellence, cannot in itself be ‘fulfilled’: yet in the perception of the fact that ‘gold is yellow’, the copula too is a given. The sentence is filled up in its entirety simultaneously with its formation at the categorial level. It is in this sense that intuition itself takes on a form. Categorial intuition, as opposed to sensory or sensible intuition, is simply the evidencing of this formal fact, which characterizes any possible intuition. I do not see-perceive primary visions and their link: I see immersion, Quine would say: I see the whole, and perceive an act of realized synthesis. This, Vailati would add, is the sense in which meanings function as the tracks guiding all possible perception. A categorial form, then, does not exist in and for itself, but is revealed and developed through its embodiments, through the concrete forms showing its necessity, and which unfold it according to specific programs that constitute, simultaneously, themselves as program-performers. It is thus meaning which has the power to produce forms, this constituting the intuition according to its categorial nature.

Category cannot be reduced to grammar because it is not outside the object. According to Husserl, we need to conceive of a type of grammar which is immanent to language, which must necessarily be the grammar of thought, of a thought which reveals itself as language in action, a language that, in its turn, constitutes itself as the Word of reality, like a linguistic corpus, i.e. a construction articulated linguistically, according to precise grammatical and semantic patterns, which gradually becomes reality. In contemporary terms, we could say that Husserl’s language in action is characterized by the fact that the origin of meaning within the organization of the complex system is merely an emergent property. What is self-organizing is the function together with its meaning, and it is in this sense that, as stated above, meaning for Husserl is able to produce form-functions.

It thus becomes clear how for Husserl form, or articulation, can be considered as precisely that, and can only be constituted as object through a formalizing abstraction. Hence the birth of a very specific intuition which can only be the result of a founding act. It is in this founding activity that the ultimate sense of categorial objectivity lies: this is the case, for instance, of mathematical evidence, which relates to the existence of a structure only insofar as it is accessed by an ordered series of operations.

Thus the actual reality of an object is not given by its immediate appearance, but by its foundation, it shows itself as something constituted through a precise act. The innate meaning of an object is that of being itself within an act of intuition. There is a moment, for example, when a circle ceases to be a circle by means of a variation procedure: it is this moment which marks the limits of its essence. Being itself identifies the very idea of intuition. To have an intuition of a sensible or abstract object means possessing it just as it is, within its self-identity, which remains stable in the presence of specific variations at both a real and possible level. The realm of intuition, in this sense, is the realm of possible fulfillments. To have intuition of an object means having it just as it is ['the thing itself']: breaking down the limits of the constraints distinguishing its quiddity. The intuition, for example, of a complex mathematical object means possessing it as itself, according to an identity which remains unaltered through all real or possible variations. An object is a fixed point within a chain-operation, and only through this chain can its meaning reveal itself.

It should be born in mind, however, that a categorial form can only be filled by an act of intuition which is itself categorized, since intuition is not an inert element. In this sense complex propositions can also be fulfilled, and indeed every aspect in a complete proposition is fulfilled. It is precisely the proposition, in all its complexity, which expresses our act of perception. A correspondence thus exists between the operations of the categorial foundation and the founded intuitions. To each act of categorial intuition a purely significant act will correspond. Where there exists a categorial form which becomes the object of intuition, perceived on an object, the object is presented to our eyes according to a new "way": precisely the way related to the form: we see the table and chair, but we can also see in the background of this perception the connection existing between these two different things, which makes them part of a unique whole. The analysis of the real nature of categorial intuition thus leads Husserl, almost by the hand, to the question of holism. But it is of course this question—the sum of the problems posed by the relationship between thought and its object—which, as we know, constitutes one of Husserl's basic points of affinity with his mentor Brentano. Brentano's slant on questions of this kind pointed the way for Husserl in his own analysis, and proved a blueprint for the development of the different stages of his research. For Brentano consciousness is always consciousness of something, and inextricably linked to the intentional reference. At the eidetic level this means that every object in general is an object for a consciousness. It is thus necessary to describe the way in which we obtain knowledge of the object, and how the object becomes an object for us.

While, then, the question of essences seemed initially to be taking Husserl towards the development of a rigorously logical science, a *mathesis universalis*, the question of intentionality then obliges him to analyse the meaning, for the subject, of the concepts used at the level of logical science. An eidetic knowledge had to be radically founded. Husserl thus proceeds along a path already partially mapped out, albeit in some cases only tentatively, by Brentano, gradually tracing an in-depth analysis of the concept of completeness before arriving at a new, more complex concept, that of organicity. Kant's category of totality, Brentano's unity of perception and judgment, and experimental research in field of Gestalt theory thus come together, at least in part, in a synthesis which is new. But further analysis of the concept of the organicity then suggests other areas of thematic investigation, in particular at the level of *Experience and Judgment*, principally regarding the concepts of substratum and dependence. At the end of his research trajectory, then, Husserl returns to the old Brentanian themes concerning the nature of the judgment, offering new keys of interpretation for the existential propositions of which Brentano had so clearly perceived the first essential theoretical contours.

It is this area of Husserl's thought which interested Quine and Putnam in the '70s and '80s, and Petitot in the '90s [2]. New sciences and conceptual relations enter the arena: e.g. the relationship between Logic and Topology, and, simultaneously, between perceptual forms and topological forms, etc. It is these main forces which shape the continuing relevance and originality of the line of thought, the secret link, as it were, between Brentano, its originator, and the two main streams represented by Husserl's logical analyses on the one hand, and the experimental research of the Gestalt theoreticians on the other.

Quine's and Putnam's revisitation of the Brentano-Husserl analysis of the relation between perception and judgment was of considerable importance in the development of contemporary philosophy. It was no isolated revisitation, however, Husserl's conception of perception-thought relations constituting a source of inspiration for many other thought-syntheses. Recent years, in particular, have witnessed another rediscovery of the phenomenological tradition of equal importance: that linked to the philosophical and "metaphysical" meditations of the great contemporary logician K. Goedel. Its importance at the present moment is perhaps even more emblematic, in comparison, for instance, with Quine's and Putnam's rediscoveries, with respect to today's revisitation of Husserlian conception. For many aspects, Goedel's rereading constitutes a particularly suitable key to pick the lock, as it were, of the innermost rooms containing Husserl's conception of the relations between perception and thought.

The departure point of Goedel's analysis is Husserl's distinction between sensory intuition and categorial intuition [3]. Goedel, however, speaks in terms not of categorial intuition but of rational perception, and it is precisely this type of perception which allows for a contact with concepts, and through which we reach mathematical awareness. In his opinion, the conceptual content of mathematical propositions has an objective character. The concepts constitute an objective reality which we can only perceive and describe, but not create. In this sense this form of rational perception is in some ways comparable with sense-perception. In both cases, according to Goedel, we come up against very precise limits, possible illusions, and a precise form of inexhaustibility.

A very clear example of this inexhaustibility is provided by the unlimited series of new arithmetic axioms that one could add to the given axioms on the basis of the incompleteness theorem: axioms which, in Goedel's opinion, are extremely self-explanatory in that they elucidate only the general content of the concept of set. Goedel's comment on this in 1964 is as follows: 'We possess something like a perception of the objects of set theory'. This perception is a sort of mathematical intuition: a rational perception. But how can the intuition of essence be reached? How is it possible to extend awareness of abstract concepts? Or to understand the relations interconnecting these concepts: i.e. the axioms which represent them?

None of this, in Goedel's opinion, can be done by introducing explicit definitions for concepts or specific demonstrations for axioms, which would necessarily require further abstract concepts and the axioms characterising them. The correct procedure is, conversely, to clarify the meaning, and this act of clarifying and distinguishing is, for Goedel, the central nucleus of the phenomenological method as delineated by Husserl. The theorems of incompleteness would seem, in effect, to suggest the existence of an intuition of mathematical essences (of a capacity in us to grasp abstract concepts), for which no reductionist explanation is possible. This kind of intuition is required above all for specific mathematical problems, for obtaining proofs of coherence for formal systems, etc. The theorems thus demonstrate clearly that that particular essence constituting "mathematical truth" is something more than a purely syntactical or mechanical concept of provability, while guaranteeing full mathematical rigour.

A rigorous science, in other words, as Husserl maintained, is more than a purely formal science. It also requires a transcendent aspect, and it is at this level that new mathematical axioms gradually come to light, arising not only from formal and deductive procedures. The unlimited series of new arithmetical axioms which present themselves in the form of Goedel's sentences, and which can be added to the already-existing axioms on the basis of the theorems of incompleteness, constitutes, in particular, a classic example of this process of successive revelation-constitution. These new axioms clearly represent precise evidence which is not extrapolable from preceding axioms via mere formal deduction. They can thus be used in order to solve previously undecidable problems. According to Goedel, this is a clearly-defined way of explaining our intuition of an essence. An even more interesting example is provided by the Paris-Harrington theorem, a genuinely-mathematical statement referring only to natural numbers which, however, remains undecidable at the PA level. Its proof requires the use of infinite sets of natural numbers, the theorem providing a sound example of Goedel's concept of the need to ascend to increasingly more elevated levels of complexity to solve lower-level problems. In a number of works between 1951 and 1956, Goedel returns to one of his favorite examples: the unlimited series of axioms of infinity in Set theory. These are not immediately evident, only becoming so in the course of the development of the mathematical construction. To understand the first transfinite axioms it is first necessary to develop the set theory to a very specific level, after which it is possible to proceed to a higher stage of awareness in which it will be possible to "see" the following axiom, and so on.

In Goedel's opinion, this is a very impressive example of the procedure of meaning clarification (as well as of the process of rational perception) Husserl had posited. It is precisely by utilizing our intuition of essence as related to the concept of a "set" that set-theoretic problems in general can be solved. It is also necessary, he goes on, for us constantly to recognize new axioms logically independent of those previously established in order to solve all mathematical-level problems, even within a very limited domain. One case in point is the possible solution of the Continuum problem.

Here Goedel states explicitly that the theorems of incompleteness demonstrate how mental procedures can prove to be substantially more powerful than mechanical same, since the procedures they use are finite but not mechanical, and able to utilize the meaning of terms. This is exactly what happens in the case of the intuition of mathematical essences. In offering us the possibility of understanding the nature of this process of categorial intuition, Phenomenology allows us to avoid both the dangers of Idealism, with its risk of an inevitable drift towards a new metaphysics, and Neopositivism's instant rejection of all possible forms of metaphysics.

While the theorems of incompleteness are not derivable from the doctrine of Phenomenology, they offer a better focus on the irreducible nature of mathematical essences, not least, for instance, through the clarification offered by the concept of 'mechanically-computable function' as analyzed by Turing. Goedel thus finds hidden truths within an epistemological perspective that many may have considered outdated and obsolete. His conceptual instruments, however, belong to an analytical tradition which is not that of Phenomenology. According to the great Austrian logician, Phenomenology is, basically, a method of research, it consists of a manifold of procedures of meaning clarification and these procedures appear indissolubly connected to specific patterns of selective activities growing up in a continuous way.

Cognitive activity is rooted in reality, but at the same time represents the necessary means whereby reality can embody itself in an objective way: i.e., in accordance with an in-depth nesting process and a surface unfolding of operational meaning. In this sense, the objectivity of reality is also proportionate to the autonomy reached by cognitive processes.

Within this conceptual framework, reference procedures thus appear as related to the modalities providing the successful constitution of the channel, of the actual link between operations of vision and thought. Such procedures ensure not only a "regimentation" or an adequate replica, but, on the contrary, the real constitution of a cognitive autonomy in accordance with the truth. A method thus emerges which is simultaneously project, *telos* and regulating activity: a code which becomes process, positing itself as the foundation of a constantly renewed synthesis between function and meaning. In this sense, reference procedures act as guide, mirror and canalisation with respect to primary information flows and involved selective forces. They also constitute a precise support for the operations which "imprison" meaning and "inscribe" the "file" considered as an autonomous generating system. In this way, they offer themselves as the actual instruments for the constant renewal of the code, for the invention and the actual articulation of an ever new incompressibility. Hence the possible definition of new axiomatic systems, new measure spaces, the real displaying of processes of continuous reorganization at the semantic level. Indeed, it is only through a complete, first-order "reduction" and a consequent non-standard second-order analysis that new incompressibility will actually manifest itself. Therefore, the reference procedures appear to be related to a process of multiplication of minds, as well as to a process of unification of meanings which finally emerges as vision *via* principles. Here also the possibility emerges of a connection between things that are seen and those that are unseen, between visual recognition of objects and thought concerning their secret interconnections. Hence, for instance, according to Boccioni: "*la traduzione in forme plastiche dei piani atmosferici che legano ed intersecano le cose*". In other words, this is the connection between the eyes of the mind and the intuitions relating to the nesting of meaning, a meaning which is progressively enclosed within generative thinking and manages to express itself completely through the body's intelligence [4].

A functional analysis of the kind reveals even more clearly, if possible, the precise awareness that, at the level of a cognitive system, in addition to processes of rational perception, we also face specific ongoing processes of semantic categorization. It is exactly when such processes unfold in a coherent and harmonious way that the "I" not only manages to emerge as an observation system, but is also moulded by the simultaneous display of the structures of intentionality. Through the intentional vision, the "I" comes to sense the Other's thought-process emerging at the level of its interiority. The drawing thus outlined, however, is meant for the Other, for the Other's autonomy, for its emerging as objectivity and action. This enables me to think of the autonomy of the Nature that "lives" (within) me.

At the level of intuition-based categorization processes, the file is selected from the ongoing morphogenesis. When the original meaning manages to express new lymph through a renewed production of forms, the self-inscribing file might express its unification potentialities through the successive individuation of concepts which, however, are selected and moulded at an intuitive level. Hence the possibility of an actual "inscription" to the same extent as the morphogenesis, but also the realization of a reduction process, the very laying down of an original creativity within a monodimensional and dynamic framework. It is exactly when the reduction is carried out, though, that the procedures of reflection, the identification of limits and completion can be performed on the basis of the constant support of the *telos*' activity, of the primary regulation activities proper to the organism, taken as ongoing projectuality.

5.2 The "Thinking I" and the Nesting of the Original Meaning

With respect to this frame of reference, Reality presents itself as a set of becoming processes characterised by the presence-irradiation of a specific body of meaning and by an inner (iterative) *compositio* of generative fluxes having an original character. These processes then gradually articulate through and in a (partially-consistent) unifying development warp with internal fluctuations of functional patterns. It is this functional, self-organizing and "irradiating" warp, in the conditions of "fragmentation" in which it appears and is reflected at the interface level through the unfolding of the canalization process, that the network-model progressively manages to reconstruct and replicate within itself as regards its specific functional aspects, ultimately synthesising and reflecting it into an operating architecture of causal programs. In this way, it is then possible to identify a whole, complex "score" which will function as the basis for the reading-reconstruction of the aforementioned functional warp. However, to read-identify-represent the score will necessarily require the contemporary discovery-hearing of the underlying harmony. Only the individual capable of representing and tuning the work as living harmony, and the score as silent object, will actually be able to depict him/herself as "I" and as subject. This individual will then not only be able to observe objects, but will itself be able to see the observing eye, modeling those objects. The I able to portray itself as such will be able to rediscover the root of the very act of seeing, positing itself as awareness and as the instrument allowing the emergence of the "thinking I", and, conjointly, of the metamorphosis of the original meaning.

It is thus through the continuous metamorphosis of the network that new Nature can begin to speak, and Reality can channel itself (*in primis* as regards the external selection), in accordance with its deep dimension, ultimately surfacing and expressing as an activity of synthetic multiplication, i.e. as a form of operating generativity at the level of surface information and as a "thinking I" able to reflect itself in (and through) the work outlined by the network-model. It is the face-texture of the effected reconstruction which provides the guidelines for the I's edification; and indeed the "thinking I" which gradually surfaces reflects itself in the constructed work, thereby allowing the effective emergence of an "observer" which reveals finally itself as a cognitive agent able to observe the Nature around him in accordance with the truth, i.e. we are actually faced with the very multiplication of the cognitive units. The system is thus able to see according to the truth insofar as it constitutes itself as an I and as consciousness, i.e. in proportion to the extent it can "see" (and portray-represent) its own eye which observes things.

In this sense vision is neither ordering, nor recognizing, nor pure comparison, nor, in general, simple replica, but is above all a reading-reconstruction of the (becoming) unity of the original body of meaning (with operating self-reflection): a process of progressive identification and assimilation of this unity in terms of an adequate texture of self-organizing programs able to portray itself as such, a process which becomes gradually autonomous and through which, *via* selection, in a renewed way and at surface level, Reality can canalize the primary modules of its own complex creative tissue: i.e. surfacing as generativity and nesting as meaning. The better the reconstruction, the more adequate and consistent the canalization: the system will function ever more sophisticatedly as a reflecting and self-organizing filter. As a matter of fact, in parallel to this an "observer" will progressively arise through the narration and the methodical verification of the distinctions relative to the functional forms managing to move at the level of the unitary and cohesive articulation of the self-organizing programs. As narration and synthesis, the I posits itself as autonomous and as the increasingly adequate mirror of a precise "metamorphosis": namely, the metamorphosis proper to an intelligent network which grows into autonomy. The mirror is image-filled at the moment of selection, when new emergence can simultaneously come about and "eyes" can then open and see both things and their meaning. The "thinking I" which surfaces and the meaning which nests thus fuse in the expression of a work which ultimately manages to articulate and unite itself with the awareness-*Cogito* and the ongoing narration: an observer thus joins a work acting as a filigree. The resulting path-*Via* can then allow real conjunction of both function and meaning. The result will be not merely simple generative principles, but self-organizing forms in action, creativity in action, and real cognitive multiplication: not a simple gestaltic restructuring, but the growth and multiplication of cognitive processes and units, i.e. the actual regeneration and multiplication of original Source according to the truth. The adequate work of unification-closure of network programs, which joins and encapsulates, at the level of the ongoing emergence and self-reflection, the selection internally operated by meaning according to the living warp-filigree, constitutes the real basis of vision in action. In actual fact it comprises a multiplicity of interconnected works, to each of which is linked a consciousness. In this way the aforementioned unification necessarily concerns the continuous weaving of a unitary consciousness, albeit within the original fragmentation of the micro-consciousness and the divided Self.

It is from this viewpoint that vision appears as necessarily related to a continuous emergence, in its turn connected primarily with the progressive articulation of a self-expressing and self-synthesizing I. As the system manages to see, it surfaces towards itself and can, then, identify and narrate itself as an "I", and specifically as an I that sees and grasps the meaning of things: in particular the emergence related to the meaning that is concerned with them. At the moment the aforementioned work becomes vision (expressing itself in its completeness), it simultaneously reveals itself as a construction in action and at the same time as the filter and the lynch-pin of a new canalization through which new Reality can reveal itself unfolding its deep creativity. Meaningful forms will then come into play, find reflection in a work, and be seen by an "I" that can thus construct itself and re-emerge, an I that can finally reveal itself as autonomous: real cognition in action.

I neither order nor regiment according to principles, nor even grasp principles, but posit myself as the instrument for their recovery and recreation, and reflect their sedimentation in my self-transformation and my self-proposing as *Cogito*. Actually, I posit my work as the mirror for the new canalization, in such a way that the new emergent work (the self-organizing mirror), if successful, can claim to be the work of an "I" which posits itself as an "added" observer. It is not the things themselves that I "see", then, but the true and new principles, i.e. the meaningful forms in action: the rules-functions linked to their emergent meanings. I thus base myself on the "word" which dictates. Hence the possibility of seeing Nature *iuxta propria principia*. The world thus perceived at the visual level is constituted not by objects or static forms, but by processes appearing imbued with meaning. As Kanizsa stated, at the visual level the line per se does not exist: only the line which enters, goes behind, divides, etc.: a line evolving according to a precise holistic context, in comparison with which function and meaning are indissolubly interlinked. The static line is in actual fact the result of a dynamic compensation of forces. Just as the meaning of words is connected with a universe of highly-dynamic functions and functional processes which operate syntheses, cancellations, integrations, etc. (a universe which can only be described in terms of symbolic dynamics), in the same way, at the level of vision, I must continuously unravel and construct schemata; must assimilate and make myself available for selection by the co-ordinated information penetrating from external reality. Lastly, I must interrelate all this with the internal selection mechanisms through a precise "journey" into the regions of intensionality.

Whoever posits himself as a tool-modality for an adequate reading process on behalf of Goddess-Muse (whoever adds, namely, himself to Nature along the execution of a score) has no choice but to listen to the melody that comes from the echoing of meaning within the Temple of life, the harmony, that is to say, linked to the possession on the part of the truth of his opening eyes. It is the self-organization of himself on behalf of the hero as a biological network (along the embodiment process) which allows the effective realization of this possession as well as the opening up to petrifaction and the self -recognition in an image, which is the reflection of himself in the face of the Goddess (only the hero that is reflected in his possessed being can, indeed, come to listen to the harmony). Here is the I of a mind in action. It is to the extent that meaning, *via* Ariadne, comes to guide and support the observation activity through the embodiment (in accordance with the truth) that original creativity can, in turn, surface to itself thus avoiding the dissipation up to determine the birth of a specific replication process: hence the possibility to achieve the life of a body as real invariance. I can posit, in this sense, myself as an I only by listening to the Other. Indeed, I can play the score which concerns me only if I am able to listen to the melody of the Other which comes to inhabit me and marks the possession of my eyes on behalf of the truth: here we can find the primary root of self-conscious I. What emerges are circuits of the flesh which, through integration and assimilation, are ultimately reflected in the half-closed eyes of Ariadne, albeit only once the precise meaning of the effected infixions has been understood. It is by a similar means that the body of the Minotaur is so smooth at the end of the metamorphosis, opening his eyes to the role of the Goddess and achieving invariance by immersing himself within her. Having discovered and manifested the trajectory of the constraints (and sutures) within himself, he is able to offer himself as support in the action of "conception". The Minotaur proffers himself as Ariadne's means of self-reading, thus of presiding over the ongoing embodiment: the means, that is to say, of unwinding the thread within the file. The Goddess-Muse who comes to be reflected in his eyes (the Minotaur finally becoming reflected in her face, as Picasso's famous litograph has it) represents in herself the achievement of meaning as meaning dislocated in ambient incompressibility, an achievement which seals the accomplishment of the metamorphosis.

It is Ariadne who carries the light with her, in her hand, and who, as truth in action and Goddess-Muse of meaning, guides and illuminates the steps of the Minotaur, of the original emotion which explodes in a thousand cascades and which wishes to possess her. To sharpen creativity and avoid its consumption by the fire of dissipation, the thread must be unraveled and Daphne-Ariadne must arborize as a labyrinth (although open to an inevitably partial exploration). Some inner light must oversee the nesting process and thereby avoid the fall into crystallization: the light which is offered the Minotaur as a guide through the labyrinth and as the secret thread relative to the plot of the infixions. It is exactly in the light of this thread that a file can then witness the opening of the Minotaur's eyes. The shepherd and the Nymph: the Nymph becomes Nature and the shepherd an observer, but only insofar as meaning self-organizes according to a precise nesting process. In this way, rather than fragmenting himself in the fire of dissipation, Apollo is obliged to anchor himself to the stone of observation. Apollo has no choice but to become embodied if he is to avoid falling victim to dissolution, but this means entering the realms of invariance by establishing an indissoluble bond with the observing eye. An incarnation for death in anticipation of salvation and the saving of life.

The thread has been unraveled and gathered into its secret luminosity, and shadow can now reveal itself in all its complexity along the wounds of light: the wounds which have turned the originally wild and radiant creature into man and martyr, through the path indicated by the wounds inflicted along the route to cognition: Narcissus-Minotaur (creativity in action) illuminated (and possessed) by new meaning even as he dies enclosed in his dream (Endymion). On the other hand, and conversely, the original light, the file, will only be able to reveal itself through the spurts which spring up along the path of the shadow: Marsyas, the incisions and the spurting blood. These incisions, together with the spurts, cause the irruption of a new file: hence the outlining of the light of an outpouring shadow, and Marsyas illuminated by new life (albeit in the Other) even within his death—the life of the poet and of chimeras. What this represents is the shadow of light in action: shadow becoming life, and light, in turn, becoming cognition: the infixions lead to the stone as the incisions lead to the gridiron. While Marsyas burns in the fire of the meadows, Narcissus drowns in the waters of meaning. He bears himself like the modality of the unraveling of the thread, like Ariadne's reading of herself, the very elucidation of herself in accordance to the truth proper to the effected observation. The eyes of the Minotaur's mind gradually open in accordance with a phantasmagoria of images and forms (written in the language of mathematics) as the eyes of Ariadne, once again in the house of meaning, narrow in the secret light shining from within themselves. The conceptual instruments of the craftsman are rarified in their turn as the eye of Clio's intellect burns brightly, the Muse having returned to the house of creativity. While the Minotaur succeeds in observing, Marsyas succeeds in thinking (through the programmes and models of the brain); his song is taken to himself by the God as Work in action and Artefact-programme (as a successful simulation machine) supporting inspiration. It is the burning gridiron and the new, self-inscribing "phrase" which feed the God's new thought and which supports the rediscovered triumph of his light: here we can find the file in action within the thread as well as the origins of the new irruption. Disembodiment takes place when Clio returns to the house of creativity, and the God becomes pure categorial and pure enthusiasm. At the end of the metamorphosis, the craftsman's brain closes in a chain-link of programmes and a soul may burn and issue forth in the fire of (intellectual) abstraction and of the Work-programme: we are dealing with a simulation which opens to new productivity, the productivity of meaning, and to a semantics which becomes generative.

Realizing the labyrinth and revealing its roots are the necessary steps if the Muse is to illuminate the original information with truth, allowing it to constitute itself as semantic information (the Minotaur as St. Sebastian) and as prelude for the new conception. It is in the moment the Muse reads herself as a score, subsequent to the full emergence of creativity as harmony linked to specific motifs and invariants, that a new conception can take place and the Painter can use his brush to depict the inborn harmony. The journey into abstraction begins here, now that it is no longer Ariadne who is present with her thread, presiding over the unraveling of the incarnation, but Clio with her file presiding over the volutes of abstraction. In the case of Ariadne we can inspect a Muse writing herself as a score while Nature unravels itself through the successive addition of different observers. The hero who adds himself as observer endorses the inscription and thus the self-reading of the Goddess-Muse; it is precisely in this sense that the opening of the labyrinth represents the means whereby Ariadne manages to read herself and unravel her own thread. If this reading reveals as successful the labyrinth will finally acquire a human nature. Lastly, the Muse will be fulfilled and will reflect the hero: eternity and invariance joined together but with respect to the ongoing replication.

The hero (Narcissus) realizes in itself as a stone (and as a column of the Temple of life) an engraving concerning a surface representation (by fixed points and symbolic retractions) of the original *Sinn* which affects him. In general, the hieroglyphics and the images at the level of the Temple concern, first of all, the writing of the names of the different heroes through the indications of the performing actions that underlie, at the semantic level, their existence. Narcissus can see only to the extent to which he can engrave his image (through the embodiment) and petrify. Moreover, insofar as he petrifies he is also obliged to drown in the waters of meaning. Only in this way, he can manage to see (and exist) as an I and as a subject. Actually, when I play as a subject the score related to my own body, enclosing myself in the presentation of this very score, a harmony necessarily comes to inhabit the eyes of my mind (and I can, finally, identify myself through this harmony). In this way, Narcissus positing himself as a stone and as the primary factor of replication, identifies himself with the eye set forever in the stone in accordance with the eternal time of a Destiny (as shown by S. Dalì), the Destiny proper to the invariance of his being. *Esse est percipere* and coincides with the unfolding of a Destiny. It is the realization, at surface level, of the body of meaning (the very reading of its image in the mirror) that gives rise to the eternal replication of Narcissus as a flower according to the truth. In this way, the hero stands as an instrument for the operational nesting and the constrained irradiation (on the surface level) of meaning as an acting unification drive. Hence the unfolding of a Nature full of meaning and necessarily linked to the successive "addition" of different observing systems. The Minotaur who, his metamorphosis completed, succeeds in reflecting himself as man and mortal is no longer a multiplicity of a thousand emotions, but a mind able to reflect within itself the meaning which concerns it and in which his mind is in turn reflected. Thus Narcissus-Minotaur can offer himself as the source of the possible replication, as a nucleus rooted deeply within life, like a hero who gives life having inscribed in himself life's *Sinn* in accordance with the truth, a hero, in particular, able to offer himself, through engraving, as witness to the achieved inscription of this very *Sinn* within his flesh. The result is a consciousness (adjunction + testimony + observation) which is always consciousness of a body and its sensations, as well as of the ideas which reach embodiment through inscription. Narcissus (as achieved embodiment) deploys (and recognizes himself in) the truth belonging to Nature as invariance (i.e. the truth related to life as invariance). Marsyas on the other hand, (as achieved abstraction), captures within himself the life of the Work as morphogenesis (i.e. the life related to meaning considered as the very root of morphogenesis). Narcissus observes his reflection in the mirror and sees the face of the Muse, recognizing himself within it and ideally engraving within himself the essentials features of this same face. He immerses himself in his own truth in consonance with the infixions perpetrated, thus succeeding in seeing objectively, although along the construction of himself (and in his offering himself as source and model of replication). What emerges is an incarnate body which manage to live on the surface and appears as perfectly honed to the specific functionality of its own being (insofar as it operates as replication in action): the functionality of a flower by means of which an Annunciation may come to manifest itself. The inscription and the testimony-consciousness therefore represent the foundation of its being as a cognitive being, a being in life but in accordance with the truth.

In the case of Marsyas, on the other hand, incisions and cipher represent the foundations of his being as an agent of transformation, of his being for truth but according to a specific embodiment. Government (through observation) and use (through manipulation): I see in that I observe, and inscribe the *Sinn* of the Other within me by offering myself as witness. It is through these means that I incarnate. At Marsyas' level, however, I think in that I palpate and manipulate (the brain palpates the world as a network in action. We are faced with Nicolais's glove in action as well as with a soul burning in the abstraction). On the one hand, truth is the necessary condition for seeing: becoming flesh implies precisely this—the emergence of creativity in truth and its offering itself as semantic creativity (at the level of categorial intuition). On the other hand, thinking is only possible in the new life: in meaning emerging into life, and offering itself as productive meaning (at the level of intuitive categorization). In other words, thinking is only possible within the Work, through the *Via* concerning the operant abstraction and Marsyas' burning in the air. In Marsyas' case it is the Other who becomes a witness within me: the consciousness of the Other which I hear/feel in my inner being and in my progression towards the necessary addition. I add myself to Nature, and am added to the Work to enable another consciousness to be born animated by the pipe which enchants. Consciousness immerses itself in Nature once, and once it exalts itself (and separates) within the Work. The testimony is once within the crystal and once within the smoke, in the volutes of a sound which curl within each other until the beat of renewed harmony can be heard in the delineation of new morphogenesis.

5.3 The Journey of the Blind in the Underworld of Language

It is possible to reflect on the nature of the self-organizing process at the biological and cognitive level—of the process, that is, governing the emergence of life—only on condition that precise reference be made both to the orderings (at the semantic level) and the imagination (at the conceptual and generative level) variously introduced on each occasion. The Minotaur will in fact manage to see in and through the truth only in so far as he is able to acquire true autonomy during the gradual realization of his detachment from the pure flames of creativity. This in its turn will come to coincide with the Goddess of meaning, Ariadne's revealing herself to him and for him as intentionality in action, thereby selecting him with the threads of Grace. In this sense the countenance of the Goddess, in its objectivity, will inevitably be consonant with the imagination in agreement with which the Minotaur will gradually bring about his own metamorphosis. Only in this way can meaning come into being in truth, and the truth reveal itself over Time; only in this way, that is, can the Goddess, pure Form in action, proffer herself as an effective measure of Time, marking it through her action and presiding over the emerging "theory". Hence the final sense of the tribute of autonomy and the offer of the "cipher": through his sacrifice the Minotaur allows the Goddess to conceive on the basis of self-reflection, albeit with prime reference to the "theory" which has come to reside in Minotaur's own mind. She observes and "reads" herself, growing on herself quite naturally (the portrait of Jacqueline by Picasso, Paris, Musée Picasso) by means of Minotaur's achieved autonomy and the fact of his facing death: '*verrà la morte ed avrà i tuoi occhi* (*death will come and will have your eyes*): (C. Pavese)'. If illuminated by the eyes of the Goddess, the shadow of the shepherd's hand in Poussin's painting "*Et in Arcadia Ego*" (Louvre, Paris) will be that of the scythe: a shadow marking achieved awareness of himself and the confines, at the level of memory too, of his own mortality: *Et in Arcadia Ego*. The Goddess will now be able to rise and become a constellation, opening up to a new brain (Marsyas) and new Time. If I address the Goddess towards different paths I will then, in this sense, give rise to new types of objectivity, thereby referring to new origins (of Time) and new calculations with regard to the entropy. What previously might have seemed entropy at its maximum could now produce very different results: verses vs. schemes, invention vs. imagination. Marsyas invents possible worlds, while the Minotaur imagines actual realities. The former burns in life, the latter drowns in truth: Eigenvalues and eigenforms. Yet new capabilities can arrive and emerge only through the extroversion of Silenus' brain. Invention is the seed for imagination, and vice versa.

Marsyas, like St. Erasmus, relates himself to a network and a brain which is everted into itself, thus connecting the natural mind to the brain at an artificial level. Just as Perseus sees the reflection of Medusa in his mind (the shield), so the Goddess (in the semblance of Medusa) sees the figure which has come to inhabit the mind of Perseus, albeit reflected in the mirror-book of stone, (the book of Nature which she is careful to consult), relocated, that is, within the framework of a language now made digital through artificial "montage". In other words, the Goddess is able to reflect and read herself in the reflection in the stone, i.e. in the mind of the Minotaur who renders himself to the stone-mirror and offers himself to the Goddess as the bringer of the cipher, thereby allowing the Goddess to read herself through reference, however, to a text which originates with him. While the Minotaur in the guise of Narcissus, following his Destiny, renders himself to the stone, the Goddess accepts the offer and reflects herself through the cipher, thereby creating the conditions for throwing off both the fixation and the Rule, i.e. the natural state assigned her by Destiny. The birth of Marsyas significantly marks the moment of liberation and opening out: the transition from Nature to the Work. As the progenitrix of Marsyas, the Goddess will, then, become his bride, entering into the world of Culture (in correspondence with the coming into being of the Assumption: A. Carracci " *Assunzione della Vergine*", S. Maria del Popolo, Rome). The Goddess of reading now possesses the cipher by means of which she can read herself, but is also able to consider herself as rule in action as she flies high in the heavens. She will then be able to change the rule in the course of the action, thus presenting herself as a new rule which changes the rules. Athena is now preparing to be born—Athena the Goddess of which Medusa represents one of her attributes, a simple shield, yet living and functional. The act of fixing (turning to stone) is in effect only one of Athena's aspects. Born of Jupiter's brain, she in turn gives birth to a brain (Marsyas), thus positing herself as the matrix of the revisable thought. Within a framework of this kind the heroes and their dance who subsequently arrive posit themselves as effective support of Nature's (coming into) being, just as Marsyas' burning can be read as support of the coming into being of the Work. The God wraps in flames the hero who offers the extroversion of his cortex, while the Goddess wraps in the waters the hero who through fixing on himself allows her to be reflected in his mind. If Marsyas burns within, inhabited and possessed by the Spirit, Narcissus in his turn offers himself to the stone (like an eye set in stone). The Spirit can provide God with nutriment in view of the new irruption, and in the Narcissus' eye the Goddess, in her turn, will be able to rediscover the cipher allowing her to conceive. The Minotaur who constitutes a fixed point for the Goddess's reflection therefore opens to subsequent self-definition in the heavens of the abstraction of new orderings following the Assumption. In order for these orderings to be applied, however, the Silenus must each time be sacrificed. The Minotaur must render himself to Nature, and Marsyas to the Work in agreement with the Spirit which feeds the flames. Methodical selection operates as a filter with regard to the growth of the cerebral volutes. If Creativity acts as a filter *via* Method, Marsyas must still preventively supply the material for extroversion, offering himself as permanent support for the Work. Only in this way can he be added as craftsman. The application of the software allows Nature to participate in the observer's detachment. The shepherd becomes vision and observer, but only in the course of his observing and rendering up to a skull (Guercino "*Et in Arcadia Ego*", Palazzo Barberini, Rome). At the level of Nature the detachment of the Minotaur (in the guise of Narcissus) involves cyclic death and resurrection. Marsyas' death by fire, on the other hand, signifies unification and eternal life, albeit with regard to the Work. Each moment in the Work is eternal, while every moment of Nature is transient. This transience, however, is in function of eternity, and vice versa. There is, in any case, a necessary evolution of the filter and of the set of orderings (meaning as *Forma formans*). The song of Apollo (or of Orpheus) cannot be continually renewed in the flames without the sacrifice of Marsyas, i.e. without deployed artifice: only through this can new creativity find its place. *Natura Naturans* is a filter constituted in itself by drawing on its network of intensities a functional ideality constructed at a cognitive level: i.e. at the level of the specific laboratory of simulation provided by Marsyas' brain. The Minotaur becomes vision in the truth, and Marsyas thought in life: through his offer of spirals he allows the God to activate the Method, thereby preparing for the irruption.

The Minotaur (in the guise of Narcissus) in his role as shepherd reflects himself (in the waters) and fixes himself to a skull, training his sight on the realization of himself following the detachment. Self-reflecting, like Narcissus he recognizes himself in the fixation: in the positing of himself for himself, as fixed point, thereby acquiring self-awareness in his own death. He drowns in the reflection, and his mind's eye will in its turn stare at him from the depths of the waters. Only during the self-reflection and his fixing on himself will the Minotaur bring about detachment in self-recognition: perception and apperception. By effecting the detachment the Minotaur reveals himself as able to offer the Goddess the cipher of his sacrifice and his self-rendering to a file, the file of a life as it enforms at the level of the ongoing self-reflection and fixation. Medusa fixes within herself all that she glances at, turning it into stone. Staring at herself in her encounter with Perseus she inevitably petrifies herself, though as a result of the duplicity. The Goddess who reveals herself to herself by means of the book-mirror of stone and the Minotaur's mind (in being a subject of fixation), and who obtains from the Minotaur (and from Narcissus) the gift of the cipher (namely, the actual secret regarding the subtraction of the Minotaur from the chaos through detachment) fixes herself, in fact, in the result of a reflection, albeit a reflection effected with regard to a "theory" and to that peculiar stone-mirror showing on its surface as it were the film related to *Natura Naturata*. Perceiving her action thus reflected on a screen she is able to recognize the orderings she herself implemented and to read them as such as the hero constitutes himself as autonomy consonant with the effected detachment. This is she who, reflecting herself in the mind of the Other, gradually confines herself within the world of Reflexivity, a world consonant with a reflection on reflection, thus enabling conception (Velasquez). This explains how the Goddess self-organizes herself together with her creativity, thereby bringing about the birth and nourishment of Bacchus—i.e. evoking the divine child's descent from the heavens. Hence derives a birth which is pure and celestial, and a path which is new while tracing the road to abstraction. While Narcissus, staring at himself fixes himself and reaches death (in the stone) but within invariance and the affirming of his own existence (thereby opening up to a Nature which is cyclical and the dance of the heroes), the Goddess stares at herself fixing herself in her orderings as they come to life in parallel with the constitution of the Minotaur's mind, and locks herself within herself, alone (and isolated) in her abandonment, at the same time succeeding in evoking the sense of a Road which is not that of incarnation but of abstraction. The point of departure is now dictated not by the intensities but by the orderings able to open towards a faculty of invention articulated by verse. The result will be given not by the detachment of the hero (the Minotaur), but by the God's adjunction of Marsyas, the newly conceived, to himself. For all this to come about, however, the Minotaur must provide an adequate cipher by means of which the Goddess can self-reflect: i.e. the inscription in the stone-memory, through which she can fix herself in invariance, in view of the opening up to the road of abstraction and conception. While the hero accepts death in the stone as linked to vision in the truth, the newly-conceived will accept death by fire but as thought in life. Through Marsyas meaning can open up to the invention and the simulation of worlds, an invention which can then be disciplined by the God through recourse to the Method, deployed not for acts of a mind (arranged schematically) but in harmony with the volutes of a brain (arranged by verses). However all this can only be accessed by constituting a universe of Reflexivity and its consequent criticism. This is the true emergence, not linked to simple chaos as existing in relation to an ongoing Reflexivity, closed within itself, but to forms of reflection which on the contrary are realized in relation to the mind of the Other and the vision in the truth on the Other's part: from the vision of truth to thought in life. Thought, through imagination and reflection, is embodied (the Nymph): the hero stares at himself and renders himself to vision in the truth and invariance. When he offers the cipher, meaning, through the implemented inventivness, opens up to abstraction: no longer a mind now but a brain which at length everts through parallel volutes as it itself comes to express in accordance with an activity of simulation by means of audacious hypotheses and adequate *experimenta*.

The Goddess who comes to reflect-read herself in the eyes of the mind of the Minotaur identifies within herself and selects the stone cathedral of Reflexivity, namely, the invariance which lives. By reflecting herself she reflects in its turn Nature in truth, and conceives the Lord of the garlands. If the Minotaur in the guise of Narcissus observes himself and fixes himself, yielding to invariance, the Goddess, on the other hand, stares at herself in the eyes of the Other and frees herself from the chains of necessity by evoking the divine child. Along the volutes of extroversion she will sketch innovative linguistic realities in the heavens of abstraction, with Marsyas yielding to thought in life: no longer schemes but verses which liberate and canalize inventiveness. On the one hand, the Spirit of the God, and on the Other the mother's womb: Nature as reign of invariance, and the Work as the forge of morphogenesis. While thought reaches invariance through reflection (in fixing itself), St. Erasmus, on the other hand, avails himself of self-extroversion to free himself in the morphogenesis (in parallel to the action performed by the Lord of the garlands). Now it is Athena who chains Medusa to herself. Medusa is reflected in the mirror, the mirror-shield belonging to Perseus. Killed and reduced to stone, she will still however have the possibility of fixing, albeit as permitted and commanded by Athena. If Medusa reflects herself in the mind of the Minotaur (the Minotaur now in the guise of Perseus), the hero can change into Marsyas while the Goddess reflecting herself in the mirror of stone may come to be assumed as a constellation, thereby giving way to the passage from the Reign of Nature to that of Culture and Work. In this lies all Athena's guile in action. While Penelope reflects Ulysses' scythe within herself and is finally able to glimpse the completion of the shroud (and to recognize herself within it, through detachment), Marsyas offers the extroversion of his own brain to the Other and programmes himself to the point of burning in it and thinking but in agreement with a life which, however, is new. While the Goddess reflects herself in Nature and renders herself to pure Grace in her conception of new language, the God programmes his selective activity in accordance with the Work, thereby allowing new thought to irrupt. Marsyas allows the God to burn through him (through the extroversion of his brain) in view of expressing his enthusiasm (an enthusiasm which is ever new), and rediscovering his own creativity in the Other. This is no longer the conception of new language, but new *Sylva* which begins to burn: in conception, the triumph of truth; in irruption, that of new life. While the Goddess reflects herself within herself in Nature, the Work everts in the God who is then able to rediscover himself in the raging of the burning bush and of the *Sylva*: natural and artificial. When thought renders itself to Nature the Goddess can reflect herself in invariance (in the dance of the heroes). When language renders itself to the Work, Marsyas can evert in morphogenesis (in the fire of the God and of creativity). New abilities will emerge, creating new mind, with reference, here, to a Time unfolding like Form in action. As sublime programmer *ante litteram*, and as subject of meaning, Goedel is able to operate a self-reading as language (that of his brain): hence undecidability, compactness, *experimentum crucis* etc. This, then, goes on to Chaitin and omega: pure randomness as the last stage of language, and complete Reflexivity ending in pure randomness; randomness read as the absence of any relation and simultaneously the presence of all relations, referred, however, to active invariance, the invariance of Form in itself, in being pure and non-analytic. Yet after the sacrifice of Marsyas the Goddess can be received in the heavens as a constellation, and the Goddess who reflects herself in the stone mirror, i.e. in the film showing on the wall of the cavern, can in effect succeed in conceiving. It will, then, be the heritage of whoever represents the fruit of the conception to shape according to the score the path of a renewed Minotaur and the presentation of the new offer. This of course is Hegel in action: what is in itself and in the heritage of Marsyas (the Idea) makes itself out of itself (and Real-Nature) in parallel with the canalization of the Minotaur, but operating on the level of the ongoing metamorphosis. In its turn it will be this canalization which allows the (continually repeated) offer of the cipher: the cipher which will posit itself as the basis on which Reflexivity enforms itself, but within the confines of a system, namely the "scientific" system related to the heritage at work. With the coming about of the conception and the Silenus' flight, further capacities will emerge, through the irruption and the devastation of the system's hearth. In this sense it will be necessary to go beyond the ancient observer, who will now have to climb onto his own shoulders, though with the help of Marsyas. The oracle and the *experimentum*, the invention of pilings and verses will all be needed, and pure vision must be set aside in favour of accepting the challenge of thought, on the basis of the ongoing dialectic between Creativity and Meaning and between vision and thought: any growth of vision is of necessity passing through Marsyas. This is what allows the first observer to climb onto his own shoulders, therein giving way to the vision of the second observer in accordance with the dialectic regarding the "relative" first devised by Skolem, which is to say that reference must be made each time to a different and more complete system-randomness. Painful apertures must be made, and blindness faced; it will be necessary to descend into the Underworld, pass through the new *Sylva*, and address the new Assumption. Where one type of entropy existed another will appear, and where a specific type of relation between events was referred to, reference must be made to a different asset (Bell vs. Einstein). Changing the abilities at issue will change the process of counting, the way we see objects, etc. Our whole "narrative" of the infinite at the level of logic and mathematics will change, and Cantor will be used to supersede Cantor as it happens at the level of Skolem Paradox. The mind thus born will emerge as different every time, and will refer particularly to new orderings as they are released by the renewed conception and subsequent Assumption. In this sense there can be no effective dialectic unless there is constant mutation of vision and thought, constant reference to language which becomes thought in dependence of the triumph of Reflexivity, and to thought which becomes language in dependence of the triumph of inventiveness. This is the journey of the blind in the Underworld of language which leads to the new vision by way of the renewed irruption, enabling new capacities and the formation of a new table of the judgments at the level of the new language thus set in action. This language will be born of the metamorphosis of thought which ultimately canalizes itself as the Minotaur in action, as a metamorphosis which succeeds in witnessing the detachment albeit in the constituting of a mind which under the influence of meaning will be able to articulate in agreement with an innovative table of the judgments, producing a Minotaur which will succeed in opening his eyes in order to read, and a Nature written in mathematical characters. From this will derive the cipher, and the mirroring of the Goddess-Rule in the mind of the Minotaur: a Minotaur who drowns in the surfacing self-image which comes to live in the digital sphere mediating through the iconic language of vision. The Goddess who mirrors herself can then permit conception to take place. Language which has rediscovered itself can approach the crossing in the abstraction which will end with the sacrifice of Marsyas. What this means is that the offer and conception are each time linked to a previous conception: an evolution-history of both meaning in action and the language it vehicles: generative Semantics in action—here is the progressive unveiling of an Ideal consonant with a targeted construction and within the effective unfolding of the dialectics between *Verum* and *Factum*. Compared with a universe of Reflexivity such as that represented by Einsteinian conception of locality, Bell's equations open, for example, new scenarios of entropy, working towards a different temporality in action: whence the move from classical Physics to Quantum Mechanics, taking into due account superposition, interference, annihilation processes, the role played by the observer etc. What to classical physicists seemed the result of mysterious actions appears very differently to scientists who plan experiments around non-standard models, oracles, etc.: whence, again, an evolutionary characterization of entropy and a definition of it which takes into account the dialectics in action between actors and observers as outlined by Prigogine.

As already stated, reality is revealed on the basis of the dialectics between incompressibility and meaning, and it is within this framework that both seeing and thinking are formed: the two senses of a single Way, creating a precise circularity marked by a series of transmutations. If a sighted Subject (the observer) is unable to uncouple from the *Sylva* through the use of the imagination, the determination operated by the Form and the intervention of the orderings offered by Form itself (which, as format, not only frames the multiple but also guides and illumines the process leading intensities to use the orderings in view of the determining of Form itself), then on an effective level there can be no autonomy (of the Subject) or Reflexivity. Without Reflexivity in action, on the other hand, there can be no conception, thus no language able to lend itself to thought, thereby allowing a new *Sylva* to come into being, and new intensities and means of synthesis at the level of Form (which, in its turn, will have undergone Assumption). If successful, the process of detachment will in itself result in configuring an observer who as achieved autonomy will profoundly intersect the landscape of emerging life. The schemes determine the Form, rendering it "by" intuitions, but functioning because light offers a thread. The light feeds and canalizes vision, the very seeing by intuitions. The thread depends on the orderings utilized each time by Form: the orderings which lead to the entry into the invisible and the microscopic. Darkness unravels, wrapping in itself the intensities and ordering them with the assistance of the threads of Grace: Virgil the classic and memory. Orderings lead to the bustling of synthesis at the level of the intuitions, revealing the rule in action: here's the weld together despite the discord of life: the order of order. The regulatory framework which is reflected within itself but in the mind and vision of the Other constitutes the last, non-analytic frame which constitutes itself at the basis of every possible intuition. This determines all activity of synthesis insofar as it connects with a Rule seemingly free of all ties and which, however, closes within itself anchored to essential randomness. This of course until both the conception and the processes of liberation come into action at the level of the Road to Abstraction. These factors taken into account, to see according to objectivity means ensuring conception and therefore the renewal of language, thereby ensuring a continuous circle between language and thought. The Minotaur, as autonomy, represents an end while simultaneously constituting a means for the perpetuation of life in truth. He underpins one of the two faces of the Road (which is both abstraction and incarnation). Insofar as the orderings come to inhabit him (developing within cognition), he will arrive at existence (*esse est percipere*) although only insofar as, in his turn, he will prove himself able to achieve autonomy, positing himself as yeast for their self-nesting and renewal. The fundamental point is the fact that intensities exist which surface as hardware but with necessary reference to the action of the imagination and the effect of the software proper to the Form. What is required is a successful natural reproduction (offering itself as invariant) which is also open to Chance in relation to the renewal and perpetuation of the cycle. The objective is not, however, invariance in itself (as it might appear in Monod from a hasty first reading) but invariance in relation to morphogenesis and viceversa. In such a framework the Road would fail to be realized in either of its senses unless eigenforms came into being and artificial-level eigenvalues were also available. The origin of the evolution concerning the software, then, occurs at the level of Nature, but that of hardware at the level of the Work. This evolution will therefore necessarily be tangled and hybrid, in answer to specific moves and formalisms of a mathematical and algorithmic nature. The result is a mathematics of Reflexivity joint with a mathematics of Inventiveness: reflexive domains on the one hand, and on the other the destruction of symmetries, simulations, oracles, *experimenta* etc.: hence the importance of the studies by Skolem, Goedel, and Turing. Goedel's approach will lead to Scott's denotational semantics and the consecration of Reflexivity as shared by L. Kauffman. Goedel is also, however, largely present in Chaitin's algorithmic meditation. The mathematics of Reflexivity constitutes the Narcissus' route to self-realization at the level of biological hardware, insofar as the hero is fulfilled as invariance which in itself intersects the occurrence of natural life. Reality canalizes itself in rendering itself to Nature immersed in Reflexivity and intersected by observers representing its own invariants. If I wish to be aware of the nature which concerns me, thereby becoming a witness and interpreter of it within my own flesh throughout the evolution in action, I must posit myself as a touchstone of invariance, and thereby succeed in seeing through the truth. I must tangibly realize my limits and those of a Reality which seems confined within itself and which, however, is still in excess: "*La vita che t'affabula è ancora troppo breve se ti contiene!*"as Montale states and as Skolem and Goedel have demonstrated. This is the necessary step from Skolem's first observer to the second: from a universe of Reflexivity to an-Other. This, however, presupposes conception, going beyond Goedel to face the problem of morphogenesis, i.e. Turing and his models of living matter. As I have had occasion to write, mathematics underpins our fibers and the workings of our imagination, yet without imagination and self-organization this same mathematics would be unable to nest within its cocoon and illumine the path of Nature in the evolutionary sense. Ideal and Real joint together at the level of natural evolution. They are the developments of software as they appear at the neural level which implement the process of abstraction which will ultimately merge with new intensities and which will allow (in conjunction with the Assumption of the Goddess) the emergence of new rules in order to prime the process concerning the emergent incarnation (the Nymph). In this sense the coming into being of the Road at the level of biological hardware is in actual fact interwoven with the development of ever new software on the basis, however, of the dialectics occurring each time between creativity and meaning.

5.4 The Metamorphoses of the Revisable Thought and the Eurydice's Dream

What constitutes the subject as such on the basis of the offering of itself by the original incompressibility (as expressed in a non-relational form, and consequent upon irruption) in view of new life and incarnation, is the ongoing "extrication" process starting from the chaotic accrual of initial capacities, a process which unravels in keeping with a trajectory allowing for the emergence of specific *eigenforms* and the consequent birth of the structures of a possible rational perception. At the level of such emergence a specific mathematical practice self-declines each time throughout the ongoing metamorphosis: a practice coinciding with an effective articulation of the different capacities at play within an evolved computational domain. Eyes then open in dependence of a specific tuning of the involved computational processes as realized on the basis of specific limitation procedures which are activated in a targeted way. As a result we are faced, at the level of the practice in action, with the gradual articulation of specific categorial intuitions. The model which ultimately reflects within itself (and satisfies) the computationally formed intuitions allows the subject, acting as its intermediary, to recognize itself as such. Narcissus recognizes himself and drowns within the half-closed eyes of Ariadne which are simultaneously possessing him from within. What surfaces on the membrane is the intentionality of the Goddess, revealing itself as the very image (the countenance) in which the hero both recognizes himself and drowns, albeit as he affirms himself as a subject who lives in the realm of truth and paves it with his life. In other words, in order to satisfy mathematical practice the model must possess (and read) from within the tissue relative to the emergent categorial intuitions and implement it in accordance with the truth. Ariadne appears on the surface of the mirror (and of the waters), in her eternal ordering. Her half-closed eyes, raised to the heavens, testify her possession of the mind of the Minotaur who is now able to unfold himself as subject, a subject which as such discovers his mortality (and observes it in his gradual morphing into the model), while attaining to the certainty of his existence as a human being through metamorphosis. The implementation of this process will produce a specific ordering in a framework characterized by a fully-achieved and seemingly absolute relationality. The formless Minotaur, pure blind accrual possessing immense potentiality, is now able to open his eyes and ultimately be mirrored in Ariadne's countenance in his achieved humanity (and mortality as shown by Picasso in a celebrated lithograph). His eyes will be possessed by the truth to the point of the full articulation of the model in its self-contained concentration. The intuitions canalizing his capacities are thus truly satisfied. This seeing in action is channelled in the new flesh now born, but articulated through forms which are of a mathematical nature. What emerges is the project of a mind (of a thought through forms) in being indissolubly linked to the progressive constitution of a self-observing eye.

Actually, at the level of FOL, a reflexive space is endowed with a non-commutative and non-associative algebraic structure. It is expandable and open to evolution over time as new processes are unfolded and new forms emerge. In a reflexive domain every entity has an eigenform, i.e. fixed points of transformations are present for all transformations of the reflexive domain. As we have seen, according to H. von Foerster (von Foerster, 1981) and L. Kauffman (Kauffman, 2003) the objects of our experience are the fixed points of operators, these operators are the structure of our perception. In this sense, we can directly consider an object A as a fixed point for the observer O: $O(A) = A$. The object is an *eigenform*. In the process of observation, we interact with ourselves and with the world to produces stabilities that become objects of our perception. Our perceptual activity, in this sense, is conditioned by the unfolding of the embodiment process and is linked to the cues offered by Narcissus to meaning in action. Then $F(J) = J \cdot J$ is said to be the eigenform for the recursion F. Every recursion has a fixed point. In such a context, every object is inherently a process and the structure of the domain as a whole comes from the relationships whose exploration constitutes the domain. As L. Kauffman affirms, any given entity acquires its properties through its relationships with everything else.

A subject can thus be born in parallel with a targeted channelling since a form-articulated universe, (as it unfolds through the canalization at work and through the involved limitation procedures), succeeds in coagulating within a single structure at the level of a surface body: the body of a Nature written in mathematical characters and expressing itself in accordance with a set of specific invariants. In the light of the exemplification introduced by Myth, a model can reveal itself as able to satisfy (and reflect) Narcissus' intuitions to the extent to which Narcissus stares at himself and self-fixes (as well as self-reflects) with respect to the model recognizing himself as a subject at the level of the image (and of the possession performed by the model) and as the bearer of a specific mathematical practice in action. Meaning (Ariadne) provides the software and operates the anchorage, while, however, opening up to the "assumption". The eyes of the Goddess progressively close at the level of the mirror just as Narcissus succeeds in fixing himself as a mortal in the image-countenance which surfaces in the mirror emerging on his very membrane: the image with which Narcissus merges as he recognizes himself in the model in accordance with (and possessed by) the truth. *Esse est percipere et percipi*. What blooms on the membrane is the in-depth intentionality of the Goddess: the surface luminosity of her software offering itself up in all the purity of its relational structure. The Goddess governs by revealing herself through her light, which regulates the essential features of the game of life and allows them to be recognized. Recognizing himself and folding himself within this very game, Narcissus is thus able to engender a possible moving beyond: on the other side, among the sharp flames of the stone book-mirror which opens, the Goddess is able to retrace the reasons underlying the evolution of her natural body to the point of engendering new conception. When the Goddess retraces the "theory" of her evolving body in the stone mirror, she is actually laying the foundations for the ensuing construction of the Work by the Lord of Garlands. It is from this unfurling towards the Spirit after the gift of the flame that the process of inner liberation is born. I merge with (and close myself within) the single structure, yet conversely free myself in giving rise to the project related to the simulation Work. On the one hand, there is the reflexive domain within which the single-structure (i.e. the model) is beginning to assume its proper dimension under the focus of the self-observing eye; on the other, the achieved delineation of a frame of simulation. In the foreground now is the hero of simulation and interweaving, and no longer the hero of reflection. The expressed intentionality sanctions the move towards the construction of a Work and the intervention of the inspiration proper to the God. In this sense the ability to channel the soul through its incarnation can be thus defined only if it is then possible to unfurl towards conception, positing oneself as he who, in the Annunciation, offers the flower born from the stone. Narcissus is fulfilled in the offering and Annunciation, and exhibits himself completely in the dance of the heroes (as shown in their paintings by N. Poussin ad S. Dalì (Poussin, "*L'Empire de Flora*", Dresda, Staatliche Kunstsammlungen; Dalì, "Metamorphosis of Narcissus", London, Tate Gallery). The channelling imposes severe limitations on the model, selectively sculpting and forging it with the flame, forcing to the surface the intentionality which nurtures it at the level of meaning in action. Just as thinking through forms is subject to the procedures of limitation throughout its anchorage, so the software of meaning in its turn undergoes specific selection. If, in effect, the rule corrects and limits emotion (G. Braque), emotion in its turn fires, selects, and frees the rule by engendering hidden potentialities, albeit in harmony with the fire of a Spirit aimed at conception. It is the cipher which liberates, determining the possible birth of rule-changing rules, (to the extent that the intentionality surfaces as productivity in action) interweaving them autonomously according to selective criteria within the framework of the individuation of a new alliance. As the soul finds its channel during anchorage in view of reflection in the model, the body is abstracted within its inner fire with the prospect of inducing the emergence of a new non-relationality yet starting from the realized model and ultimately superseding it. In other words, Ariadne through conception opens up to the road of simulation leading finally to the "splitting". The type of relationality and non-relationality involved must be identified with regard to conception and irruption respectively: each time we shall be directed towards the horizons of a specific subjectivity: that of an observer or of a creative craftsman. Such are the means employed by Eurydice at the level of the exploration of the non-standard procedures, in view of identifying the type of non-relationality soon to enter the scene: that particular non-relationality which initially seems absolute (albeit anchored to the clues offered by Bacchus) but which will then be identified as relative by the very definition of the model and the consequent reflection. It is only in the light of this achieved awareness and the related ongoing selection that it will be possible the opening to the return of the dazzling light of the Lord of Garlands. The non-relationality which is revealed as relative opens to a relationality which conversely appears as absolute, as would be the case with an observer closed within his/her own universe. A new Way to abstraction will thus open up at the level of the subsequent conception. When new incompressibility is then declined in accordance with the notes of Pan, the Nymph's eyes will open in the flesh with reference to the action of a meaning retrieved in itself through the trajectory relative to the assumption process. It is from the model born of the very rib of meaning at play that those particular forms of possession capturing the eyes of the Nymph will, then, be gradually deployed allowing the Nymph to see through forms and intuitions. This is the possession which, markedly, will then allow the Nymph to enter into meaning albeit as a subject, that is to say as an observer joining the harmony of the Temple. A subject who is able to recognize himself as such on account of this very act of adjoining, of his coming to be illuminated through the playing out of the game of life and through the exercise of the possession of behalf of the truth: a possession which will signally permit ostension and ensure invariance at the level of the herm. The embedding of the self-observing eye exactly follows from the reached invariance.

The problem arises in identifying the role played respectively by the reflexive domains and by the simulation frames. It is one thing to consider a domain which is expressed through reference to FOL, through which specific structures of perception are formed, operating in the dimension proper, for example, to cellular automata or the phenomena of emergence and reflexivity which occur at their level (a dimension in which the single structure enforms itself in accordance with a purity expressed in Narcissus' case in precisely the characters of his self-closure within cyclical eternity): it is, however, a very different matter to refer to frames living at the level of second-order structures in accordance with the principles of Henkin semantics. In the latter case, we will necessarily be dealing with the edification of non-standard models and the subsequent intervention of very specific limitation procedures connected to the constituting of a self-observing eye anchored, it will appear, not only to a specific and circumscribed version of the dialectics between creativity and meaning but also to the changes induced by the continuous unfolding of such dialectics. Meaning will now assume a very specific role. The Nymph who is born will each time be the bearer of a renewed emotion, and her opening eyes will turn above all to the remains of Marsyas, the hero who initiates life, although through his cognitive work. Here we find a process of self-organization in action, but equally the need for necessary and continuous changes with respect to the involved semantics, as well as for recourse to the intervention of a different kind of selection. Being inspired means tripping oneself up and playing oracle to oneself at the level of the *experimentum crucis* through the exploration of the non-standard, and in accordance with the "fulfillment" of the model. The self-observing eye which is newly born each time reveals itself as bound to the selection made by Eurydice, the exploring of Hades, and the role that infinite numbers, for example, will come to play in the landscape of the shadows. It is therefore the above-mentioned "fulfillment" (the application of the single structure or, in Picasso's case, of the crosshatching of Ariadne's face and her half-closed eyes) which allows us to open up to that exploration and that sacrifice, both connected, from which emerges the very possibility of the new song and the re-invention of the instrument (the Saraceni's violin or Sacchi's spinet). When inspiration flows at the level of the song of Orpheus then a new Nymph will be able to arise from the notes that pour from the musical instrument, and with the Nymph a renewed process of incarnation. The creation of the song, moreover, will inevitably correspond with a process of "assumption" of meaning in action, conjointly with the definition of new software. From this derives the importance of the fundamental change operated by Henkin at the level of the General Semantics, and his ability to position himself so as truly to listen to the plot of the different selection activities. We are nonetheless aware that from this also derives the central role played by the work carried out by Tennenbaum, Kaye etc. at the level of the ongoing research on this intriguing topics. The new theorems developed by these mathematicians brought about a new awareness as regards to the more sophisticated order types which are becoming increasingly prominent and acting at the most profound level of the ongoing mathematical research. At the moment, new limitation procedures manage to graft on the trunk of such significant developments, even permitting the realization of new types of models in which the new observers who are born cannot but enclose and recognize themselves. In this sense, the procedures cited above then emerge as being in a close relationship of derivation with the questioning of the oracle: the fruit not of spontaneous invention but of a real *experimentum* on Marsyas' part which passes through his very sacrifice, ensuring the targeted flow of a renewed inspiration.

He who had enabled the God to rise again will feel within himself (albeit in the transmutation) the cries of the new flesh and the development of a new language, and will recognize his ancient remains. He will be able to see from the outside the ancient world "in relation to" the observer (in which, as such, he had recognized himself) to the extent to which he had emerged from that same world by climbing on his own shoulders. It is in the new flesh that is born (the Nymph) and in her Eros that a new mathematics will be enabled to reveal itself consumedly. Similarly it is through the limitation procedures and the work of anchorage brought into play by meaning that the new emergent structures will be able to be recognized by an 'I' that constitutes itself through its coming to be reflected in the intentionality which succeeds in possessing it. Hence the observer in action, the self-observing eye which looks at Nature, but through forms, and self-reflects within it while fully aware of its meaning, the meaning which has come to govern the hero. Narcissus drowns in the meaning, inundated with its light and pervaded by the truth. He encloses himself within the mathematical structures which regulate and underscore his cognitive activities. At the level of the truth he appears to coincide with the coming to be traced of the Way, with a pure incompressibility in action (and the relative axiomatic set) which has been made flesh. As we have just said, Narcissus recognizes himself in the waters and drowns paving the Goddess with life and becoming illuminated with truth, but he also recognizes himself in the intentionality proper to the Goddess which comes to shine on his own membrane. This is the Way for the hero to render himself, as realized "I", to the stone and to the dance of heroes. Intentionality gleams within him and the game in action thus begins to make its sense manifest. Narcissus identifies himself with self-control, obtaining the seal of truth from this identification (Gödel), and drowns, while simultaneously opening to the new conception and the new second-order enformation (Henkin). At the point where the initial incompressibility is fixed in all its purity, to the extent of being illuminated in accordance with the figures of a renewed invariance, there the new journey can begin. In this lies the sense of the discovery of the "relative" and of the consequent need for new exploration. The hero who recognizes himself in the achieved paving and in the emerging intentionality is able to observe the Nature in which he is immersed as a corner-stone, and can thus listen to his own song as included inside a living harmony finally perceiving within himself the holistic unity of the Temple. He will be able to pave the body of meaning with life, insofar as he comes to be possessed by it. Meaning, for its part, once opportunely forged becomes a model possessing through truth and becoming paved with life as it reflects within itself the hero who offers the cipher. The observer who adds himself in this way morphs into the herm and approaches invariance while simultaneously allowing meaning to start down the way of abstraction. Meaning and its software then allow the drawing (and the "fixing") of emergent information, thereby making it wholly intentional and determining it as the matrix of new life by means of the design of renewed figures of invariance. The result is the surfacing on the membrane of a countenance in which the observer finds self-recognition. Invariance that at the moment of its activation appears as absolute cannot but reveal itself, then, as relative, thereby preparing the ground for trasmutation. For his part Marsyas, burning on the upturned cross, is unable to escape feeling all the absolute of fresh irruption on his own skin, thus preparing the way for a new incarnation. The observer who paves the body of the Goddess with his own life recognizes himself in the beauty, coherence, and completion of the mathematical "lines" which come to fix themselves on the surface, in the figures of invariance, that is to say, which finally come to shape the body of Nature which welcomes him and engulfs him in its meanders. By the light of the moon the lines proper to the game of life manifest themselves in all their wondering radiance: Narcissus can self-identify in (and with) them, hence the coming into being of the herm. His life can be revealed in this language cognate with a realized and manifest invariance, and with his life the very sense of the game which he will play out (albeit in accordance with an emerging self-awareness). In the possession exercised by meaning in action he is enabled to recognize the secret seal of life which concerns him and which identifies life as such (by charging it with the due meaning). It is in this sense that the model necessarily individuates itself by means of the conception and the entrance onto the scene of the Lord of Garlands. Meaning now has the possibility of offering itself as productive. At Narcissus's level what occurs is a relational symphony (which, as it were, "lives" him), realized within him and bearing in itself the key to the intentionality of the Other—hence the "phrase" (M. Proust) which appends its seal. The observing eye cannot but self-observe and thus proceed to examine its relationship with Nature in which it is immersed and which offers the necessary support to this self-observation. It is only through reference to the Temple that the column is able to fix itself through Medusa. When I fix myself through Medusa I listen to the web of invariants echoing in me, the web in which I drown and in which I recognize myself but at the level of a Nature that welcomes me into itself (although in accordance with the government expressed by meaning at play). In fixing myself I see the Goddess's countenance surfacing on my membrane; I am now wholly part of Nature as process (and functional corporeity) in action. When the procedures relative to control are established the interrogation can begin. What is produced then is no longer a body, a game animated by rules and invariants (the game of life), illuminated by the Goddess of meaning, but an activity of analysis and simulation, albeit in the invention of the new machine and the passing of Method as it will be envisaged in view of the definition of the set-up regarding the new game. There derives a simulation activity which penetrates the meanders of an unknown mathematics in order to find new, non-standard models, and perform the necessary *experimentum crucis* (but on one's own skin): an *experimentum* which may provide or indicate the necessary hints in view of a possible construction of new patterns of life.

The problem lies in the fact that my choices are not made on the grounds of the sole application of criteria rooted exclusively in my interiority; it is God-emanated inspiration which actually engraves me with its tracks, predicated on Marsyas' challenge, Icarus' mad flight, and enquiry conducted on his own skin by the Silenus (becoming a Spinx to himself and consequently dying on the inverted cross). Inspiration is prominent in indicating the way, which lies in my own passing. A new incarnation can then ascend from my flayed flesh, but to a melody fresh with innovation. Through this incarnation a renewed mathematics can be embodied, and a new, unheard Nature may appear and speak in mathematical characters linked each time to an emergent conceptual revolution. The Nature of Galileo's stars is very different from that of Cézanne, the latter a Nature of a thousand emotions constituted of the Mountain of reflexivity and enchantment, crossed by streams causing it to palpitate and grow in and on itself, and subjected to continuous metamorphoses. In Cézanne a new mathematics emerges through the incarnation produced through (and on) completion of the Work: an incarnation through which it is possible to recognize oneself as the new Minotaur and through which the mathematics which 'rises' (U. Boccioni, "*La città che sale*", New York, The Museum of Modern Art) encloses us, like Narcissus, at surface level. The model which comes to read him from within is what sanctions him to recognize himself as a subject which computes, albeit through his being satisfied with respect to the model itself. When this happens the new mathematics explored at the splitting level comes to be part of the hero's actual fibers, and surfaces on his membrane as the actual structure of the software of meaning in action. A similar confluence comprises the hero's gift of the flower which heralds conception. With the new listening, the Nymph will slowly open her eyes, detaching herself from Silenus' very skin; Cézanne's card-players (P. Cezanne, *Le Jouers de cartes*, Paris, Musée d'Orsay), for their part, will be able to move autonomously in space proportionately to the achieved individuation of the "cypher" related to the model. They open up to the new vision (and are its witnesses), at the same time losing themselves within it in view of the new conception. Thus Titian, like Midas, bears witness to Silenus' sacrifice and at length presents the new Bacchus child (Titian, "*The torture of Marsyas*", Praga). The Silenus (Marsyas-Titian), will be able to recognize himself in the intended model, albeit at the cost of the transmutation to follow, his very enclosure, like Narcissus, in the image concerning the ongoing petrifaction. When the Nymph turns her gaze on Titian the model is already there, and Titian is aware of the passing working within him, and acts in accordance with operations linked with the new, emergent Method and the new means of concept construction. These new means will accompany him to Hades during the Bacchus child's renewed entry onto the scene. In this sense choices come to light proportionately to my offering myself as new vision, as a self-observing eye opening to the world on the basis, however, of the newly-implemented limitation procedures: choices which are *necessarily* revealed in proportion to a specific degree of awareness at the biological level. Rather than simply using new apparatus, during my flaying I make myself an instrument: the choices can thus come into being through me, yet in the Other. The new Nature which is born, newly clothed with fresh vestments, will bathe me with the eyes of the fleshly puppet able to observe the old vestiges through detachment. It will rise to the eminence of he who embodies a story, its witness: who opens the eyes of the flesh and incorporates with Nature as observer. As the God of simulation rises again and turns his eyes on its mute remains through the ensuing incarnation as new language begins to animate him from within, Nature's new body is gradually constituted mathematically, through recourse to specific limitation procedures and new software. The possibility is thus created for a self-observing eye to add itself to Nature, albeit with reflection and self-awareness, the eye of the Nymph guiding the process of growth and adjunction which concerns her. The observer is reflected in an ideal body which, opening up to a reading of itself and the annunciation, can finally bring about conception. The Lord of Garlands then born will be able to drain the inspiration into himself, bringing about a renewed irruption with, however, the recovery of the scent of a primordial creativity. As stated above, it is at the moment of subsequent petrifaction that a new mathematics of invariance will enter the scene and speak in the fibres of the observer commensurate with the achieved reflexivity and apperception, producing the possibility of a closure within a renewed intended model which will inevitably prove relative since commensurate with the objectively drained information. It is not, then, simply by resorting to further apparatus or new glasses that we shall supersede the achieved level of petrifaction: to obtain new instruments we must necessarily undergo the interrogation of the oracle and the journey in the shadowy footsteps of Eurydice: Marsyas' renewed sacrifice, then, and the necessary death of the Father as for example in von Trier's film Nymphomaniac. Cézanne, like Titian, self-encloses and attains truth in the girl's eyes among the trees crossed through by the inspiration with respect to which he proffers himself as medium, thereby opening up to new song. From this song and these notes will be born the vision which, self-modelling, will guide the Nymph to her enclosure within the image of herself, finally positing herself as sculpted stone (in the waters) able to capture within itself, albeit on the surface, a mute creation self-rendered to eternity (Bernini, Memorial to Maria Raggi, Roma, Santa Maria sopra Minerva). The Nymph received life from the Work of the artist, on the wave of the rediscovered splitting: it is this which allows Silenus' transmutation. The incarnate mathematics is able to reveal itself as both system and model, particularly as the pluridimensional inlay of harmonious proportions which in Cézanne, on the wave of the anchorage operated by meaning in action, ultimately succeeds in beating out Nature's rhythm in view of concentrating wholly on the raising of a body, on the addition to Nature of a self-perceiving system, of a subject, that is to say, which finally stands in all the complexity of its biological equilibrium and self-awareness, and which renders new Nature permanently visible in itself in an act of ostension. The system dictates and self-inscribes via rules: Nature appears as mathematically inscribed since the system fulfils itself as a model and a Temple (harmony) through the self-constituting of the observer, the very inlay of the self-observing eye. The model emerges keeping with Narcissus' adjunction, in the eternity of its beauty and the fire of truth. A lunar light shines out, high in the heavens: that which presides over the mathematical fulfillment of the game of life. New conception can then follow. Control has been achieved and conception will have made possible a renewed plot as well as a new interrogation of the Sphinx. With the coming of inspiration the selection will start to vary, as posited by Henkin, and a new level of the many-sorted case will appear, hence the new validity of Löwenheim-Skolem theorem and the fulfillment of the new incarnation. This in its turn brings about a renewed dialectic between standard and intended; for this to take place, however, the act of flaying will necessarily have occurred, like the journey into the Underworld and the passing of the new Method.

Eurydice, living among shades, fails to "see": on the wave of the splitting she explores worlds conceptually, thinks *via* concepts, and delineates extensions (at the level of mathematical invention) in view of a possible new incarnation ensuing the new song which burst forth. With the birth of the Nymph new, specific forms of recursion and ordering will be possible, linked to a renewed many-sorted version, with new computational practices (at the mental level) which, however, will employ new methods and which will emerge from the chaos ensuing devastation, also in dependence of the surfacing of new language. The result will be the birth and consolidation of a new invariance, with new methods and new individuals at play. When, in fact, new fixed points will emerge at the co-evolutionary (and infinitary) level on the basis of new limitation procedures, we will witness new eyes in the flesh, and new Nature, as well as the actual establishment of new intended models, albeit in reference to more complex computational practices than hitherto. We will have been worn by new glasses consonant with more adequate systems of numbers, and will be able to find again the ancient natural numbers, and operate with them in accordance, however, with a more extensive savor. At the moment I am able to clarify (and extend) the nature and the functionality of the intended model, and refer it to a precise mathematical practice related to a process of renewed incarnation, I will be able to calculate in accordance with increased confidence to the point of "seeing" natural numbers more appropriately within their confines and throughout their actions, allowing me to locate even more precisely the relational properties at play, albeit not in an absolute sense given that the achieved individuation still sanctions an opening towards conception and simulation, the renewed exploration of non-standard models. The road to the new vision and alliance begins at this point, here we can find the true sense of morphogenesis. New systems of numbers begin to be visible, and will wear the garments of a new Nature emerging. The old relations will appear as even more "fixed" in their nucleus, yet charged with a functionality and meaning never hitherto expressed. This is the case in Chardin (but also in the late Titian): in the former's case a new mathematics of chaos is about to come into play, its related methods, as exposed by the artist on his own pulse and skin, sanctioning the emergence of properties at variance with the ancient fixed points which now come to be immersed in a more sophisticated information flux. This in turn institutes a complex dynamics involving the conceptual thought of the artist as architect and agent, the irruption of chaos, the necessary design for specific trajectories, concatenated symmetry breakings, etc. There will no longer be the presence of simple domains of reflection, procedures of limitation and cellular automata (as characterized in FOL), but in their place "flights" and trajectories, turbines of light, and superimpositions of dimensions occasioned, however, by opening from within, with the consequent necessary devastation and percolation already underway. Chardin's plums seems more real than that painted by artists who preceded him, exposing their *qualia* with apparent immediacy, but the result of the painter's work constitutes the point of arrival of a complex itinerary. With the all-obliterating flaying and devastation new invariants are constituted, and it is through reference to such invariants and to Eurydice's work that new languages and new instruments will gradually become visible. The artist will emerge from the chaos proportionately to his ability to control it, he exploits its immense potential, and proceeds to the obligatory interrogation of the oracle. Self-similarity and meaning as use are able to produce a new reflexivity at the level of a new incarnation, in its turn able to obtain, in accordance with a renewed many-sorted version, precise indications as to the system of cuts which of their own accord will shape the newly-born reflexive domains with their load of intentional information. The domains will thus be incardinated within a story albeit in response to a precise *bricolage* construction. The Nymph who opens her eyes speaks a language consonant with these cuts, conforming herself to them and to the new productivity of meaning in action which will be matched by corresponding new software at the level of the universe of rules.

In other words, the theorems now arising, for instance, at the level of the ongoing research concerning the order types, give no direct indication about the right state at the level of the many-sorted case, but open to the draining away of inspiration, possibly sanctioning the creation of new instrumentation at the level of our brain. The opening to non-standard procedures and the identification of new order types should, therefore, be seen in relation to the emergence of a new set of primordial capacities and their intriguing development as well as to the intervention of the reading operated by the model within the process of constitution of the mind of the observer along the ongoing composition of the machine-languages at stake. A new conception of the mathematical practice comes, therefore, to take shape. When we identify, for example, the limits and then enlarge the horizon to the non-standard case and infinitary structures, we are no longer faced only with formulas, equations and limitation procedures. The eigenforms which are to be born identify new modalities of perception, raising the stakes for a new vision and allowing us to come to be "forked", at the developmental level, from eyeglasses that enhance our cognitive growth (working from inside). In this way we are able to discover the presence of different computational layers evolving directly at the level of our own being and determining our own cognitive activity. New mathematics will come to emerge at the level of our own fibers, a mathematics of which we will then have to trace the roots placing us throughout the transmutation as oracles to ourselves. In this way we achieve a harmony in which we identify ourselves, but in view of a renewed openness at the semantic level: new worlds will open up in connection with the ongoing change of paradigms. We will be revealed, thereby, as participants in a story which is our own story and which identifies what we truly are, a story which materializes, as G. B. Vico says, by means of our own work, but as humans beings, thus becoming History: here is Clio and her enthusiasm.

References

1. Husserl, E. (1964). *Erfahrung und Urteil*. Hamburg.
2. Petitot, J. (2008). *Neurogéometrie de la vision* (p. 397). Paris: Les Editions de l'Ecole Polytechnique.
3. Gödel, K. (1986). *Collected Works* (Vol. 1). In S. Feferman et al. (Eds.), New York: Oxford University Press; Gödel, K. (1972). On an extension of finitary mathematics which has not yet been used. In S. Feferman et al. (Eds.) (1989) Kurt Gödel. *Collected Works* (Vol. II). Oxford.
4. Carsetti, A. (2004). The embodied meaning. In A. Carsetti (Ed.), *Seeing, thinking and knowing* (pp. 307–331), Dordrecht: Kluwer Academic Publishers; Carsetti, A. (2010). Eigenforms, natural self-organization and morphogenesis. *La Nuova Critica*, 55–56, 75–99; Carsetti, A. (2011). The emergence of meaning at the co-evolutionary level: An epistemological approach. https://doi.org/10.1016/j.amc.2011.08.039.

Bibliography

1. Anderson, P. W. (1985). Suggested model for prebiotic evolution: The use of Chaos. *Proceedings of the National Academy of Sciences of the United States of America* 3386.
2. Atlan, H. (1992). Self-organizing networks: Weak, strong and intentional, the role of their underdetermination. *La Nuova Critica, 19–20,* 51–71.
3. Ayala, F. J. (1999). Adaptation and novelty: Teleological explanations in evolutionary biology. *History and Philosophy of the Life Sciences, 21,* 3–33.
4. Bais, F. A., & Doyne Farmer, J. (2008). The physics of information. In P. Adrians & J. Van Benthem (Eds.), *Philosophy of information* (pp. 609–684). Amsterdam.
5. Bell, J. L., & Slomson, A. B. (1989). *Models and ultraproducts.* Amsterdam.
6. Benacerraf, P. (1965). What numbers could not be. *Philosophy Review, 74,* 47–73.
7. Beggs, E., Costa, J. F., & Tucker, J. W. (2010). Physical oracles: The turing machine and the wheatstone bridge. *Studia Logica* 35–54.
8. Berget, S. M., et al. (1977). Spliced segments at the 5′ adenovirus 2 late mRNA. *Proceedings of the National Academy of Sciences, 74,* 3171–3175.
9. Birkhoff, G. (1967). *Lattice theory* (3rd ed.). Princeton.
10. Boolos, G., & Jeffrey, R. (1989). *Computability and logic* (3rd ed.). Cambridge.
11. Boolos, G. (1975). On second-order logic. *The Journal of Philosophy, 72,* 509–527.
12. Brooks, R. D., & Wiley, F. O. (1986). *Evolution as entropy.* Chicago.
13. Bray, D. (1995). Protein molecules as computational elements in living cells. *Nature, 376,* 7–31.
14. Carnap, R. (1956). *Meaning and necessity* (2nd ed.). Chicago.
15. Carnap, R., & Bar Hillel, Y. (1950). *An outline of a theory of semantic information.* Technical report. N.247, M.I.T.
16. Carnap, R., & Jeffrey, R. (1971). *Studies in inductive logic and probability.* Berkeley.
17. Carsetti, A. (1992). Meaning and complexity: A non-standard approach. *La Nuova Critica, 19–20,* 109–127.
18. Carsetti, A. (1993). Meaning and complexity: The role of non-standard models. *La Nuova Critica, 22,* 57–86.
19. Carsetti, A., (Ed.). (1999). *Functional models of cognition. Self-organizing dynamics and semantic structures in cognitive systems.* Kluwer A. P.: Dordrecht.
20. Carsetti, A. (2004). The embodied meaning. In A. Carsetti (Ed.), *Seeing, thinking and knowing* (pp. 307–331). Kluwer A.P.: Dordrecht.
21. Carsetti, A. (2010). Eigenforms, natural self-organization and morphogenesis. *La Nuova Critica, 55–56,* 75–99.
22. Carsetti, A. (2011). The emergence of meaning at the co-evolutionary level: An epistemological approach. https://doi.org/10.1016/j.amc.2011.08.039.
23. Chaitin, G. (1987). *Algorithmic information theory.* Cambridge.

A. Carsetti, *Metabiology*, Studies in Applied Philosophy, Epistemology and Rational Ethics 50, https://doi.org/10.1007/978-3-030-32718-7

24. Chaitin, G., & Calude, C. (1999). Mathematics/randomness everywhere. *Nature, 400,* 319–320.
25. Chaitin G. (2007). *Meta maths: The quest for omega*. New York.
26. Chaitin, G. (2010). Metaphysics, metamathematics and metabiology. *APA News, 10*(1), 11.
27. Chaitin, G. (2013). *Proving Darwin: Making biology mathematical*. New York.
28. Church, A. (1956). *Introduction to mathematical logic*. Princeton.
29. Cooper, S. B. (2009). Emergence as a computability-theoretic phenomenon. *Applied Mathematics and Computation, 215*(4), 1351–1360.
30. Cresswell, M. J. (1973). *Logics and languages*. London.
31. Crutchfield, J. P., et al. (1986). Chaos. *Scientific American, 255,* 38–49.
32. Davies (Ed.). (1998). *The new physics*. London.
33. Dean, W. (2013). Models and computability. *Philosophia Mathematica (III), 22*(2), 51–65.
34. Delahaye, J. P. (1989). Chaitin's equation: An extension of Gödel's theorem. *Notices of the A.M.S., 36,* 984–987.
35. Denbigh, K. G., & Denbigh, J. S. (1985). *Entropy in relation to incomplete knowledge*. Cambridge.
36. Di Giorgio, N. (2010). *Non-standard models of arithmetic: A philosophical and historical perspective* (M.Sc. thesis). University of Amsterdam. https://www.illc.uva.nl//MoL-2010-05.text.pdf.
37. Dougherty, E. R., & Bittner, M. L. (2010). Causality, randomness, intelligibility, and the epistemology of the cell. *Current Genomics, 11*(4), 221–237.
38. Feferman, S., et al. (Eds.). (1986, 1990, 1995). *Kurt Gödel: Collected works* (Vols. I, II, III). Oxford.
39. Gaifman, H. (2000). What Gödel's incompleteness result does and does not show. *The Journal of Philosophy, 9708,* 462–470.
40. Gibbs, J. W. (1902). *Elementary principles in statistical mechanics*. New Haven.
41. Gillies, D. A. (1973). *An objective theory of probability*. London.
42. Gitman, V. (2015). An introduction to nonstandard models of arithmetic. In *Analysis, logic and physics seminar,* April 24. Virginia Commonwealth University.
43. Goedel, K. (1986, 1990, 1995). Collected Works, I, II, II (S. L. Feferman et al. eds.): Oxford.
44. Goldblatt, R. L. (1976). Metamathematics of modal logic. *Reports on Mathematical Logic, 3,* 19–35.
45. Goldman, S. (1953). *Information theory*. New Jersey: Englewood Cliffs.
46. Grassberger, P. (1984). Information aspects of strange attractors. In J. Pergand et al. (Eds.), *Proceedings of NATO Workshop on Chaos in Astrophysics*. Florida.
47. Grossberg, S. (2000). Linking mind to brain: The mathematics of biological intelligence. *Notices of AMS* 1358–1374.
48. Halmos P. (1963). *Lectures on boolean algebras*. Princeton.
49. Harrington, L.A., Morley, M. D., Scederov, A., & Simpson, S. G. (Eds.). (1985). *Harvey friedman research on the foundations of mathematics*. Amsterdam.
50. Henkin, L. (1953). Banishing the rule of substitution for functional variables. *The Journal of Symbolic Logic, 18*(3), 20–38.
51. Henkin, L. (1950). Completeness in the theory of types. *The Journal of Symbolic Logic, 15,* 81–91.
52. Herken, R. (Ed.). (1988). *The universal turing machine. A half century survey*. Oxford.
53. Hernandez-Orozco, S., Kiani, N. A., & Zenil, H. (2018). Algorithmically probable mutations reproduce aspects of evolution, such as convergence rate, genetic memory and modularity. *Royal Society Open Science, 5,* 180399.
54. Hintikka, J. (1963). Distributive normal forms in first-order logic. In J. M. Crossley & M. A. Dummett (Eds.) *Formal systems and recursive functions* (pp. 47–90). Amsterdam.
55. Hintikka, J. (1970). Surface information and depth information. In J. Hintikka & P. Suppes (Eds.), *Information and inference* (pp. 298–330). Dordrecht.
56. Husserl, E. (1964). *Erfahrung und urteil*. Hamburg.

57. Hutchinson, J. (1981). Fractals and self-similarity. *Indiana University Mathematics Journal, 30,* 713–747.
58. Jaynes, E. T. (1957). Information theory and statistical mechanics (I and II). *Physical Review, 106*(4), 620–630 and *108*(2), 171–190.
59. Jaynes, E. T. (1965). Gibbs versus Boltzmann entropies. *American Journal of Physics, 33,* 391.
60. Kauffman, L. H. (2003). Eigenforms—Objects as tokens for eigenbehaviours. *Cybernetics and Human Knowing, 10*(3–4), 73–89.
61. Kauffman, L. H. (2009). Reflexivity and eigenforms. *Constructivist Foundations, 4*(3), 120–131.
62. Kauffman, L. H. (2010). Eigenforms and reflexivity. *Constructivist Foundations, 12*(3), 250–258.
63. Kauffman, L. H. (2018). Mathematical themes of Francisco Varela. *La Nuova Critica, 65–66,* 72–83.
64. Kauffman, S. A. (1993). *The origins of order*. New York.
65. Kauffman, S. A., Logan, R. K., Este, R., Goebel, R., Hobill, G., & Shmulevich, I. (2008). Propagating organization: An enquiry. *Biology and Philosophy, 23,* 34–45.
66. Keenan, E., & Faltz, L. (1985). *Boolean semantics for natural language*. Dordrecht.
67. Kohonen, R. (1984). *Self-organization and Associative memories*. Berlin.
68. Kolmogorov, N. (1968). Logical basis for information theory and probability theory. *IEEE Transaction IT, 14*(5), 662–664.
69. Kripke, S. (1975). Outline of a theory of truth. *The Journal of Philosophy, 19,* 690–716.
70. Landsberg, P. T. (1978). *Thermodynamics and statistical mechanics*. London.
71. Leinfellner, W. (2000). The role of creativity and randomizers in societal human conflict and problem solving. *La Nuova Critica, 36,* 1–27.
72. Machover, M. (1996). *Set theory, logic and their limitations*. Cambridge.
73. Manzano, M. (1996). *Extensions of first-order Logic*. Cambridge.
74. Martin-Delgado, M. A. (2011). On quantum effects in a theory of biological evolution. arXiv: 1109.0383v1 [quant-ph] 2 Sep 2011.
75. Maynard Smith, J. (2000). The concept of information in Biology. *Philosophy of Science, 67,* 177–194.
76. Mayr, E. (2001). *What evolution is*. New York.
77. Merleau, P. M. (1958). *Phenomenology of perception*. New York: Routledge.
78. Montague, R. (1974). *Formal philosophy*. Newhaven.
79. Németi, I. (1981). Non-standard dynamic logic. In D. Kozen (Ed.), *Logics of programs.* London.
80. Nicolis, G. (1989). Physics in far-from-equilibrium systems and self-organisation. In P. Davies (Ed.), *The new physics*. London.
81. Nicolis, G., & Prigogine, I. (1989). *Exploring complexity*. New York.
82. Piaget, J. (1967). *Biologie et connaissance*. Paris.
83. Prigogine, I. (1980). *From being to becoming*. San Francisco.
84. Prigogine, I. (1993). *Les lois du Chaos*. Paris.
85. Putnam, H. (1965). Trial and error predicate and the solution to a problem of Mostowski. *Journal of Symbolic Logic*, 30.
86. Paris, J., & Harrington, L. (1977). A mathematical incompleteness in peano arithmetic. In J. Barwise (Ed.), *Handbook of mathematical logic* (pp. 1133–1142). Amsterdam.
87. Partee, B. (Ed.). (1976). *Montague grammar*. New York.
88. Putnam, H. (1983). *Representation and reality*. Cambridge.
89. Quine, W. V. (1960). *Word and object*. Cambridge.
90. Quine, W. V. (1993). *La Poursuite de la Verité*. Paris.
91. Petitot, J. (2008). *Neurogéometrie de la vision* (p. 397). Paris: Les Editions de l'Ecole Polytechnique.
92. Popper, K. (1959). *Logic of scientific discovery*. London.
93. Rantala, V. (1975). Urn models a new kind of non-standard model for first-order logic. *Journal of Philosophy Logic, 4*, 455–474.

94. Reeb, G. (1979). *L'analyse non-standard, vieille de soixante ans?*. Strasbourg: I.R.M.A.
95. Rogers, R. (1971). *Mathematical logic and formalized theories*. Amsterdam.
96. Schmuker, D. et al. (2000). *Drosophila Dscam* is an axon guidance receptor exhibiting extraordinary molecular diversity. *Cell, 101*, 671–684.
97. Scott, D. (1980). Relating theories of the lambda calculus. In P. Seldin & R. Hindley (Eds.), *To H.B. curry: Essays on combinatory logic, lambda calculus and formalism* (pp. 403–450). New York: Academic Press.
98. Segerberg, K. (1971). An essay in classical modal logic. *Filosofiska Studier, 13*.
99. Shapiro, S. (1987). Principles of reflection and second-order logic. *Journal of Philosophical Logic, 16*, 309–333.
100. Shapiro, C. (1991). *Foundations without foundationalism. A case for second-order logic*. New York.
101. Skolem, T. (1933). Uber die Unmöglichkeit einer volistandigen Charakterisierung der Zahlenreihe mittels eines endlichen Axiomensystems. In *Norsk Matematisk Forening, Skrifter* (pp. 73–82). Reprinted in (Skolem, 1970, pp. 345–354).
102. Skolem, T. (1941). Sur la portée du théorème de Löwenheim-Skolem. In *Les Entretiens de Zurich* (pp. 25–47). Reprinted in (Skolem 1970, pp. 455–482).
103. Skolem, T. (1955). Peano's axioms and models of arithmetic. In *Mathematical in-terpretation of formal systems* (pp. 1–14). Amsterdam.
104. Skolem, T. (1970). *Selected works in logic*. In J. E. Fenstad (Ed.). Oslo: Universitets for laget.
105. Solomonoff, R. (1978). Complexity-based induction systems: Comparisons and convergence theorem. *IEEE Transactions on Information Theory, 24*, 422–432.
106. Talmy, L. (2000). *Toward a cognitive semantics*. Cambridge.
107. Tichy, P. (1971). A new approach to intensional analysis. *Nous, 5*, 275–298.
108. Tieszen, R. (1994). Mathematical realism and gödel's incompleteness theorem. *Philosophia Mathematica, 3*, 177–220.
109. Tolman, R. C. (1938). *The principles of statistical mechanics*. Oxford.
110. Turing, A. M. (1952). The chemical basis of morphogenesis. *Philosophical Transactions of the Royal Society of London, 237*(641), 32–72.
111. Van Benthem, J., & Doets, K. (1983). Higher-order logic. In D. I. Gabbay & F. Guenthner (Eds.), *Handbook of philosophical logic I* (pp. 275–329). Dordrecht.
112. Van Dalen, D. (1983). *Logic and structure*. Berlin.
113. Van Fraassen, B. C. (1971). *Formal semantics and logic*. London.
114. Van Lambalgen, M. (1989). Algorithmic information theory. *Journal of Symbolic Logic, 54*, 1389–1400.
115. von Foerster, H. (1981). Objects: Tokens for (eigen-) behaviors. In *Observing systems, the systems, inquiry series* (pp. 274–285). Salinas, CA: Intersystems Publications.
116. von Wright, G. H. (1957). *Logical studies*. London.
117. Varela, F. (1982). Self-organization: Beyond the appearances and into the mechanism. *La Nuova Critica, 64*, 31–51.
118. Varela, F. (1975). A calculus for self-reference. *International Journal of General Systems*
119. Varela, F., Thompson, E., & Rosch, E. (1991). *The embodied mind*. London: The MIT Press.
120. Wang, H. (1974). *From mathematics to philosophy*. New York.
121. Wang, H., & Rosser, J. B. (1950). Non-standard models for formal logics. *The Journal of Symbolic Logic, 15*(2), 113–129.
122. Wang, H. (1996). Skolem and Gödel. *Nordic Journal of Philosophical Logic, 1*(2), 119–132.
123. Wicken, J. S. (1987). *Evolution, thermodynamics, and information*. New York.
124. Willis, D. (1970). Computational complexity and probability constructions. *Journal of the ACM, 17*(2), 241–259.
125. Wuketits, F. M. (1992). Self-organization, complexity and the emergence of human consciousness. *La Nuova Critica, 19–20*, 89–109.

Author Index

A. Carsetti, *Metabiology*, Studies in Applied Philosophy, Epistemology and Rational Ethics 50, https://doi.org/10.1007/978-3-030-32718-7

Subject Index

A. Carsetti, *Metabiology*, Studies in Applied Philosophy, Epistemology and Rational Ethics 50, https://doi.org/10.1007/978-3-030-32718-7

Printed in the United States
By Bookmasters